Robert Griffin

Challenger 1 Main Battle Tank

The Tankman's Experience Continues

vol. I

KAGERO

Introduction

The Challenger Main Battle Tank (for it was only ever referred to as the Challenger 1 after its replacement, the Challenger 2, began to be seriously developed) had a relatively short career compared to the Chieftain and Centurion which had preceded it. Before the trial by fire it underwent in the 1991 war to liberate Kuwait (Operation Granby), many critics condemned the Challenger design next to the Leopard 2 and the M1 Abrams, both of which had more modern fire control systems and outperformed the Challenger at CAT87. As Robert Griffin here describes, the reasons for the Challenger's selection as the spear point of the Royal Armoured Corps in the British Army of the Rhine are complex. The Challenger embodied many compromises and was affected by political factors as much as any new British weapon system of its day.

The Challenger was a Main Battle Tank designed to foreign requirements in the 1970s and adapted to British service as a cheaper and quicker alternative to developing an indigenous replacement for the Chieftain. In this first volume of two dedicated to the Challenger 1 from Robert Griffin, the content has been carefully focused on the development and anatomy of this heavily armoured second generation MBT. The technological evolution of British tank design in the 1970s in which the Challenger originated is carefully summarized. Kagero Publishing has given this second Photosniper title by Robert Griffin a new format which we hope will be of benefit to AFV historians, modellers and former crewmen alike. Volume 2, which describes the service life of the Challenger 1 will follow shortly.

M.P. Robinson, editor.

Challenger 1 Main Battle Tank. The Tankman's Experience Continues. Vol. I • Robert Griffin • First edition • LUBLIN 2014

 • **ISBN 978-83-62878-95-6**

Editors: **M.P. Robinson, Tomasz Basarabowicz** • Photo credits: **The author would like to thank the following individuals and organizations for the use of their photos and for their help in finding suitable photos for this project: Martin Jones, Martin Cassidy, Kim Hills, Phil Richards, W. Dunbar, David Brown, John 74, Steve Cleator, Phil Harvey, Neil Bolton, The British Army, MoD, Dave Thomas, Don Lawson, Derek Whitehouse, Royal Ordnance Factory, Plain Military (Andy Brend), Dennis Lunn, Keith Frape, S. Park, Malcolm Cleverley and of course Simon Dunstan** • Drawings in scales: **Piotr Boczoń** • Design: **KAGERO STUDIO, Marcin Wachowicz, Łukasz Maj**

Oficyna Wydawnicza KAGERO
Akacjowa 100, Turka, os. Borek, 20-258 Lublin 62, Poland, phone/fax: (+48) **81 501 21 05**
www.kagero.pl • e-mail: kagero@kagero.pl, marketing@kagero.pl
w w w . k a g e r o . p l
Distribution: **KAGERO Publishing Sp. z o.o.**

The revolutionary FV4211, which bore a passing resemblance to the later Challenger 1, was the result of a design study to apply Chobham armour over a dedicated turret and hull design employing welded aluminium armour instead of steel for the basic structure. The FV4211 was otherwise completed from Chieftain components in 1971, and was the first tank designed to carry Chobham armour [Martin Jones].

Finding a Successor to the Chieftain

In 1966 the British had at last started to replace the venerable Centurion with their first Main Battle Tank, the Chieftain. The Chieftain would soldier on as a gun tank until 1994 and achieve reasonable overseas sales before being replaced by Challenger 1. It sounds very simple enough but the birth of Challenger 1 was anything but simple, for apart from all the usual criteria that affect tank design, politics probably have sometimes more influence than most, it was certainly the case for Challenger 1.

Once a new tank enters service and the vehicle procurement program is running well, design priorities switch to its replacement. This was especially true during the Cold War, and much as the 1950s had yielded technological innovations in many types of weapon systems, the 1960s had also seen advances of cardinal importance to armoured fighting vehicle design.

The Chieftain entered service in 1966 and was slated for replacement by 1985, a service life it would exceed by nine years. In the case of the Chieftain the choice of a replacement was complicated by a weak national economy and repeated defence cuts. Much of the research necessary to develop a replacement consisted of paper studies while the Chieftain itself went through its lengthy teething troubles. One such study survives from February 1976 (conducted by the Ministry of Defence and carried out by two civilians, and a major from the Royal Artillery). It lists the options for a possible Chieftain replacement for use in the 1st British Corps area in West Germany in the 1980s. The following suggestions were listed as serious options for a new British Main Battle Tank.

Code Name	Description of Option
CHA	Continued production of Chieftain Mk5 based MBT.
CHAE	Chieftain Mk.5 with Rolls Royce CV12 engine
CHAG	Chieftain Mk.5 production with new 110mm gun
CHAEG	Chieftain Mk.5 production with CV12 engine and 110mm gun
CHAN	New production of improved Chieftain Mk.5
CHB	Chieftain "E" steel hull, limited Chobham armour, V12 engine and 120mm gun
4222	FV4222, a proposed MBT with aluminium hull, anti-HEAT armour, V12 engine and 110mm gun based on FV4211
UK1	Semi fixed gun 60 ton turretless FMBT with CV16 engine, Chobham armour and 110mm gun
UK2	Turreted 60 ton MBT with CV12 engine, Chobham armour and 110mm gun.

It must be borne in mind that these were only paper suggestions (although the FV4222 was an updated specification for the FV4211, a design worthy of discussion). The report did not really reach any fine decisions and was a paper exercise, but the options it listed show clearly that by the mid 1970's a great deal had transpired in the world of tank technology, developments that had given awareness of the need for a Chieftain replacement some urgency.

The FV4211

It also shows that as late as 1976 the Ministry of Defence was far from specifying its new MBT by even basic design parameters. It is likely that with Royal Ordnance Factory Leeds busy with Chieftain production, a certain amount of consideration was devoted to keeping the replacement's general layout along the same general design parameters. This is a potential explanation for Chieftain Mk.5 derived options in the 1976 list. The principal technological development that was not included on the Chieftain series was Chobham armour, an innovation developed in 1964 when the Chieftain design was already finalised and in production. Alternately known as codename Burlington

While series production did not result, the *Aluminium Chieftain's* designed level of protection and its general layout set the trend for most of the design work performed on British MBTs for the ensuing 20 years (with the exception of the turretless FMBT). The FV4211's armour layout influenced foreign designs like the M1 Abrams. The use of major assemblies of aluminium armour on such a heavy AFV was a technological dead end however [Martin Jones].

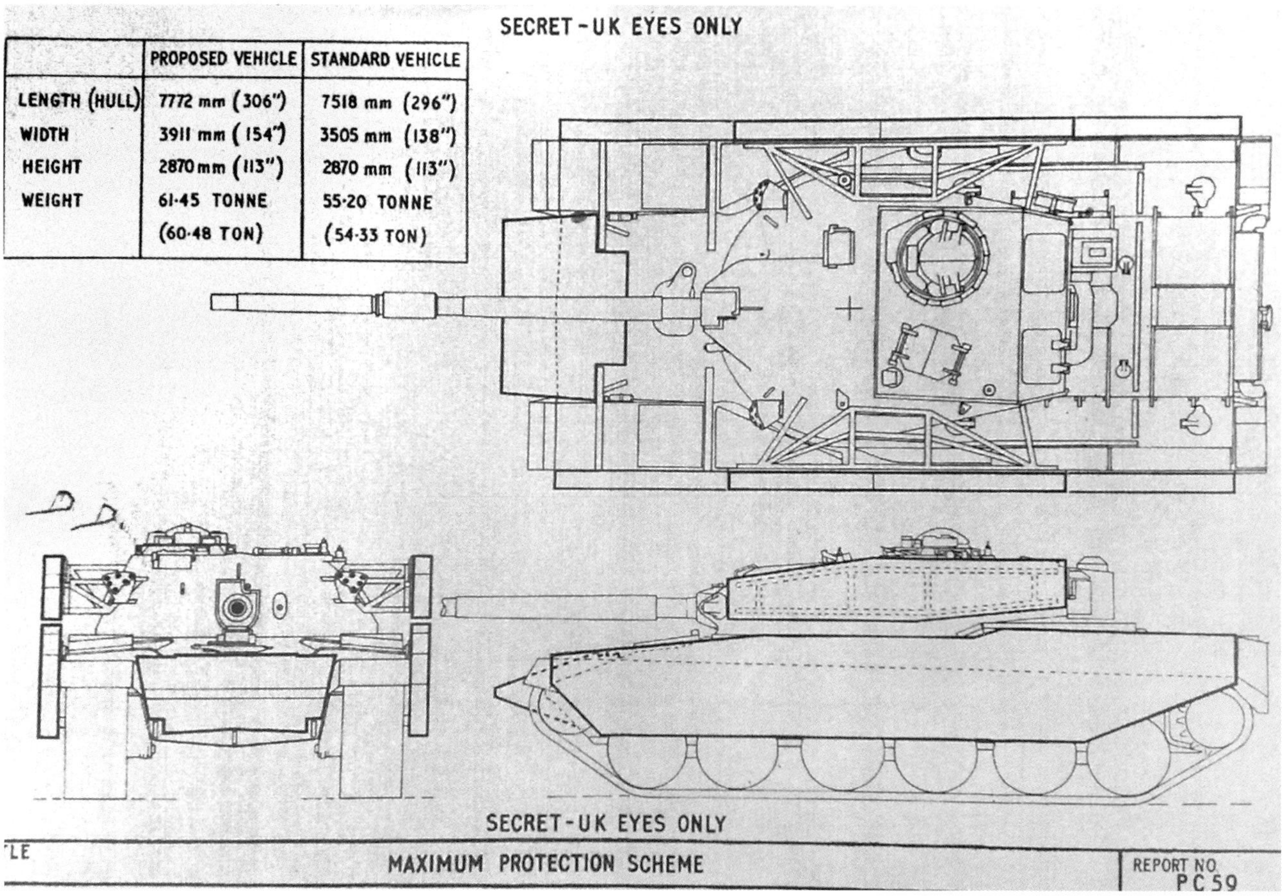

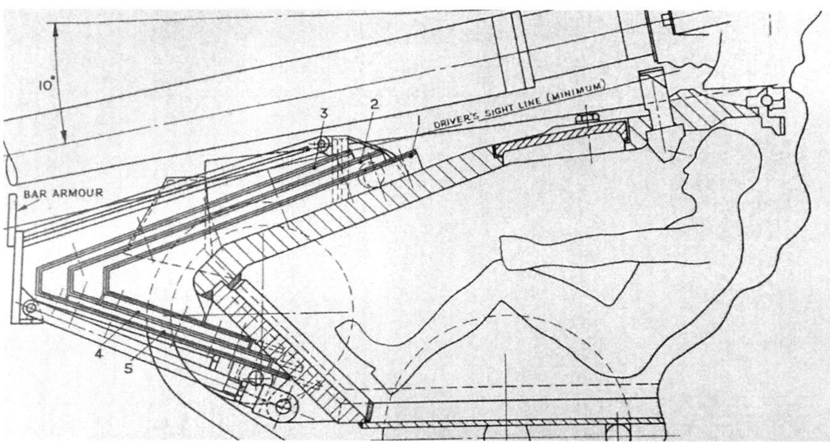

Chobham Armour was evaluated in several proposed layouts as add-on protection to the Chieftain MBT. The earliest designs were oriented towards protection from ATGM and RPG attack from the sides in a series of flat pannels reminiscent of wartime Schurzen fitted to German panzers (and similar arrangements were later seen on the Warrior IFV and CRARRV recovery vehicle).
Once test firing established that the Chobham system was equally effective against kinetic energy rounds, frontal protection became the focus of the various designs examined to protect an MBT. These diagrams date from the late 1960s [Martin Cassidy].

and codename Dorchester, two versions of this armour were in advanced stages of development and ready to test fit to an MBT by 1969. Chobham Armour was a quantum leap forward in the protection of armoured vehicles, and remains the optimal means of protecting an MBT even 40 years later. The potential advantage for a British MBT protected with Chobham armour in a tank versus tank engagement was plain to see.

In 1970 the Fighting Vehicle Research and Development Establishment (FVRDE) constructed an experimental tank based on Chieftain tank components at ROF Barnbow, the first AFV to use the new armour. This tank, the FV4211, mentioned in the list above, was a truly innovative departure from an orthodox battle tank of its generation, representing in many ways the first modern MBT as we currently define it. The FV4211 was often known as the *Aluminium Chieftain*, and it was built using available components with a composite welded aluminium hull and turret. The use of aluminium was very much in vogue for the construction of lighter vehicles at the time and the FV4211 was the first MBT built on this basis. A welded aluminium hull and turret offered advantages in both in cost and weight. It was of a similar layout to the Chieftain both inside and out and utilised the same 120mm gun and 7.62 GPMG machine gun armament (and employed the L60/TN12 powerpack).

It also was equipped at some time with the No. 21 cupola that was at one time destined to upgrade the Chieftain fleet (but which was never adopted). The vehicle was completed in 1971 and after testing was used extensively as a sales platform for Chobham armour. The FV4211's frontal arc was protected by thick panels of Chobham armour on the glacis plate and turret front, offering unparalleled protection against HEAT and armour piercing munitions. The features embodied in the FV4211 became the basis for concepts for a Chieftain replacement, although the basic design itself was too deeply rooted in the Chieftain design to adopt without changes. It would be the great expense of the new armour that would prove the greatest obstacle to its procurement by the British Army for several years to come.

The year 1971 also saw the UK tank industry receive its biggest orders since the 1950's heyday of the Centurion with

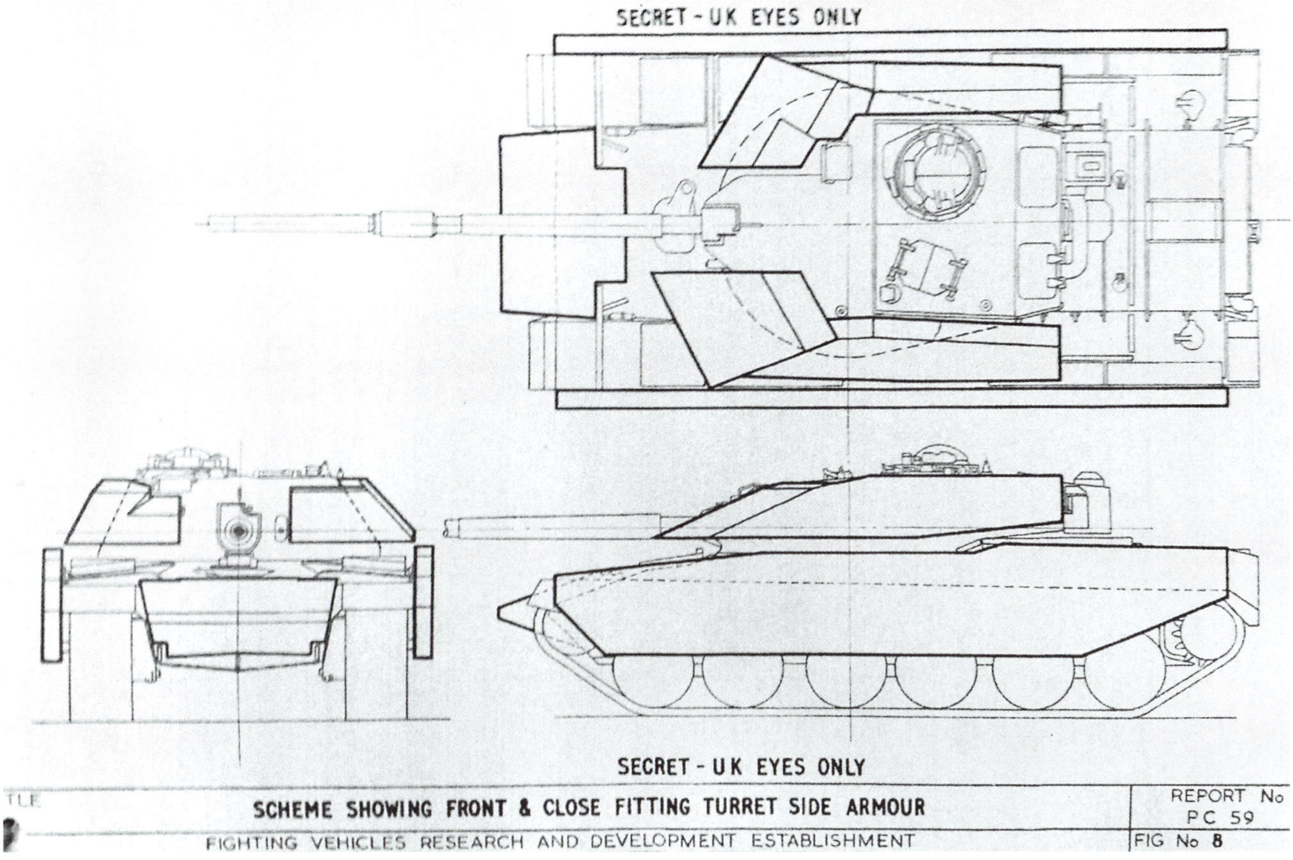

The Chieftain Mk.5/2 was the misleading designation given to the design study that was first conducted in the late 1960s and continued through the early 1970s to examine ways to implement a Chobham armour upgrade to the existing British Army Chieftain fleet by means of a retrofit program. The upgrades studied incorporated several proposals for frontal and lateral armour arrays combined with spaced armour and bar armour. The layout proposed for the Chieftain Mk.5/2 appears extremely bulky because of the complex shape of the underlying castings and the requirement of the Chobham system to be installed in flat blocks. The version pictured proposed the most complete protection of several designs. The upgrade was never proceeded with due to cost but led to the purpose built FV4211 design [Martin Cassidy].

The Concept Test Rig (or CTR); which was possibly the only Anglo-German Future Main Battle Tank program prototype that evolved beyond a paper design [Plain Military].

Remembered more often as the *JagdChieftain*, the turretless layout emulated some contemporary designs, such as the influential Stridsvagn 103 S-Tank [Plain Military].

It can be presumed that had the turretless JagdChieftain design progressed to the stage of a production vehicle, L30 gun, the CV-12 engine, frontal and lateral Chobham armour protection and a traversable hunter-killer fire control layout would have been features of the British version [Plain Military].

Interestingly, vehicles like the Stridsvagn 103 employed a far smaller length to width ratio in the track base and a far more complex mechanical layout to tackle some of the inherent problems in a turretless design [Plain Military].

The CTR was likely not the only design considered for the FMBT project with West Germany, but Anglo-German cooperation on MBT design ceased in 1977 [Plain Military].

an order for 780 Chieftain Mk.5 based tanks for the Imperial Iranian Army these would consist of 707 Gun tanks and the balance consisting of support vehicles such as the ARV and AVLB. The gun tank vehicles supplied were designated Mk.5/3P (with the P standing for Persia). They were similar to the standard UK Chieftain Mk5, but with an automatic gearbox to simplify driver training. The Iranians were very pleased with the Chieftain's general design but had reservations about the troublesome L60 power pack, and this was to be reflected in their next order.

The Anglo-German FMBT Program

Meanwhile in the search for a Chieftain replacement the UK entered into an agreement with the Federal Republic of Germany (FRG), and the two countries agreed to work on a design to be known as the *Future Main Battle Tank* (FMBT), which would replace both the Chieftain and the Leopard. Germany had already been part of the unsuccessful MBT70 program with the USA and the nascent Leopard 2 design was available to them as a fall-back, so one can imagine that the degree of necessary

The CTR *JagdChieftain*: its resemblance to the tank hunters from the Second World War is obvious. The mock- up gun can be seen across the back of the vehicle decks and the mounting bolts are clearly visible on the front glacis plate [Robert Griffin].

A head on view of the *Jagd-Chieftain*, where its Chieftain lineage can also clearly be seen. Although only a concept model, it was one several ideas for the Chieftain replacement. One novel approach utilised for this prototype vehicle was that the engine bay could accept either the L60 power pack or the German MTU unit, which many would have wished the Chieftain had employed [Robert Griffin].

The *Shir Iran* in its definitive form. Known as the Shir 2 to the British, was arguably the best combination of gun power and protection at the time it first left the factory. The M1 was still armed with the L7 105mm gun and the Leopard 2 had far less effective armour at the time. The Shir 2 however had a fire control system only incrementally better than the Chieftain series it was based upon [Simon Dunstan].

commitment to the partnership may not have been as great as may have been desired by the British. Problems quickly arose on different interpretations of each countries' requirements, but design work began and proceeded, and several designs were produced, some of which were probably too advanced for the time.

Although much that surrounded the Anglo-German program remains secret, one interesting design proposal prototype does survive in the Tank Museum at Bovington. Although it probably had an official designation other than CTR (*Concept Test Rig*), due to its shape it was and remains known as the "JagdChieftain"; due to its resemblance to the German tank killers of WW2. The decision to adopt a turretless configuration was probably influenced by the British Army's testing of the Swedish Stridsvagn 103 S-Tank and by German experience

Four Shir 2 prototypes were built for Iran prior to the Challenger program, and one at least of these received the Hydrogas suspension [Simon Dunstan].

The fate of the Shir 2s. Located on Slab Common at the REME training area at Borden, Hampshire is this prototype Shir 2 minus its Chobham armour. Still in use as a recovery aid, visible is the original Horstman suspension as found on Chieftain, fitted here with lightweight road wheels [PM].

The Shir 2 turret looks like a welded Chieftain turret and the weld marks are where the support brackets for the armour were welded. Beneath its Chobham armour the Challenger 1's own turret looks quite similar [PM].

The Shir 2 hull with the frontal Chobham armour removed, giving it a very unusual look, especially around the area of the drivers hatch [PM].

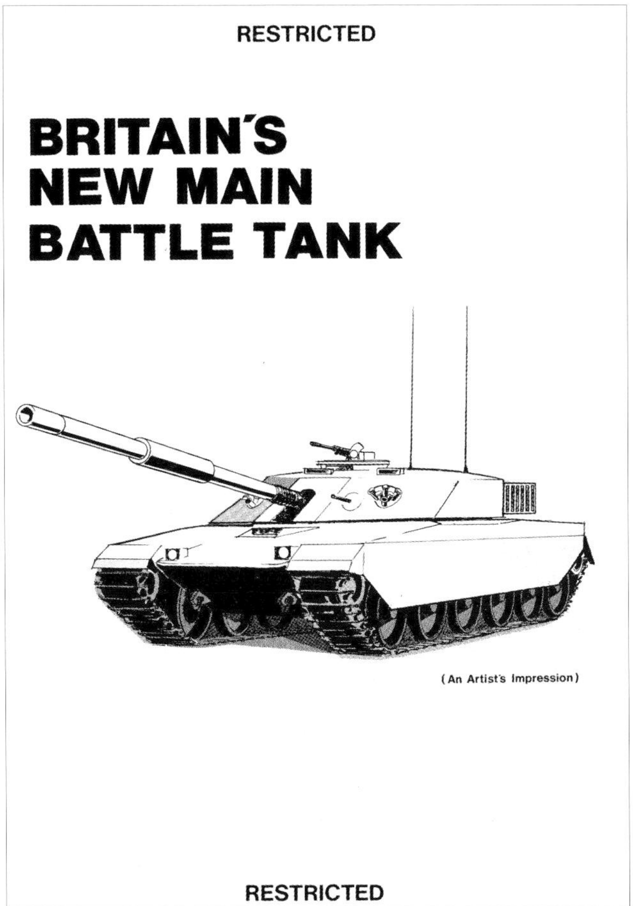

The front cover of the Ministry of Defence-produced brochure that was meant to herald in the era of MBT80. Notice how Shir 2/ Challenger-like the MBT-80 looks in the artists' impression. Inside the brochure could be found various glowing and overoptimistic statements. This brochure was issued to virtually all armoured regiment personnel in 1978 [Robert Griffin].

with turretless AFVs like the Jagdpanzer Kanone. The CTR was constructed from standard Chieftain parts and was designed to have a 120mm gun in the sloping front plate, which was specified to be protected by a massive slab of Chobham armour. A standard length Chieftain hull was employed as the basis of the CTR, which might have performed better with a shorter track base.

The CTR's gun only had movement in elevation, so traverse was achieved by turning the whole vehicle in the target direction, which proved far less efficient than hoped for in tracking tests performed at Woolwich. The vehicle was configured to be powered either by the UK L60/TN12 power pack or the German MTU power pack (and doubtless would have employed the CV12 option had it been considered seriously for production in Britain). Although joint cooperation did not produce a Anglo-German Standardpanzer in the end, it did result in the decision that Britain's future tank must follow a conventional turreted layout with a four man crew. The Anglo-German partnership did not endure, and with all the diverging doctrinal differences clouding the design, the project was cancelled by agreement of both countries in 1977. The designation FMBT did not drop

The sole survivor of the MBT80 project. The Shir 2-based Automotive Test Rig 2 or ATR 2, built to test technology for the projected development phase of the MBT80. The turret on ATR2 has long been a source of controversy. Some sources incorrectly state that it is an incomplete MBT80 turret (which is incorrect). The turret fitted to the ATR2 appears to be based on that fitted to the FV4211 prototype (less its Chobham armour) and is probably a relic left from the cancelled FV4211 testing program. It features an offset gun mounting and wider mantlet than the Shir turrets, which is very evident in the picture. The hull is very much like the Shir 2 in layout but features an aluminium rear end, which may have been an idea considered as a weight saving measure. No MBT80 prototypes were ever produced, and the projected costs of the MBT80 led to its cancellation at the paper stage and to the procurement of the Challenger 1. It is likely that the MBT80 hull would have been similar to the Shir-2, but the turret would be more difficult to speculate about with any certainty [NB].

The turret mounted on ATR2. This turret was not the basis for what would have been the MBT80's turret and is more likely left from the FV4211 program, and is preserved minus its Chobham armour. Of note is how far forward the area for the commander's cupola seems to be, with very little room for the gunner's sight. Forward of the loaders hatch is the circular location that would have housed the commander's independent sight with limited traverse [NB].

there, and was continued as a generic term for a Chieftain replacement for a few years yet however.

The Shir Program

While the Anglo-German FMBT project had been running, the Iranian Army had been shown the FV4211 prototype and had been explained the potential benefits of an MBT fitted with Chobham armour. The FV4211 had impressed them very much and was obviously a quantum leap ahead of their fleet of Chieftains, so in December 1974 they placed another large order of 1200 more Chieftains, but of an improved version, including the replacement of the troublesome L60 power pack. Development of a new power pack was given to Rolls Royce based on the CV12 diesel and David Brown was contracted to develop the new gearbox. Once these two components had been married together to form a power pack it was realised that in its present configuration, the existing Chieftain hull was not large enough. A new hull was to be developed for it, which led to the creation of the FV4030 specification, the basic design being partly based on the FV4211 but at the Iranian insistence it was to be built

The first Challenger 1 prototypes were not vastly different from the last Shir 2 prototypes [MoD].

The turret modifications necessary for the fitting of the TOGS thermal gunnery system were applied later in the trials programme (on later prototypes) [MoD].

Much use of the prototypes photos from public displays in 1980 was made in the military press of the day, which generated a misleading perception of the Challenger's capabilities [MoD].

While the Challenger offered proven gunnery and fire control systems, the power train required extensive testing to perfect [Kim Hills].

While the accepted production figures for the Challenger 1 were 409 gun tanks and 11 prototypes and development vehicles, British Army VRN numbers for 444 vehicles exist, which means that an additional 24 vehicles may have been built [Kim Hills].

The prototypes did not all carry Chobham armour. In a departure from the rather sober schemes carried by most Challenger 1s, Prototype 06 SP 40 is seen here at Lulworth in 1980 in a striking black and NATO green striped scheme [Kim Hills].

Lulworth 1980: 06 SP 40 lacking turret front stowage baskets with a Scorpion 90 visible in the background [Kim Hills].

The prototypes were extensively tested but did not achieve the level of reliability required for the power train (especially the TN37 automatic transmission) before late 1981 [Simon Dunstan].

The prototypes were kept around long after the Challenger's introduction to service: the most recent use of a Challenger prototype's turret was as a gunnery test mounting for the Swiss RUAG CTG 120mm smoothbore gun being considered as a replacement in the Jordanian Al-Hussein fleet in the early 2000s [Simon Dunstan].

Prototype 06 SP 42 being put through its paces, probably in 1981. Bovington currently has prototype 05 SP 52 on display, with earlier turret features being very much in evidence [Simon Dunstan].

with conventional armour steel for hull and turret. Long term trials had proven that the technology for the steel/aluminium hull initially proposed was not advanced enough and cracks had been discovered in FV4211's aluminium plates. This decision ultimately led to the demise of the FV4211 project, and no further all-aluminium MBT hulls have been built in Britain, with only the single prototype at Bovington left to tell the tale.

While the design of the new Iranian tank was moving ahead a batch of 150 Chieftains were produced with automatic gear boxes, improved mine protection and increased fuel capacity, designated FV4030/1. The FV4030/2 scheduled to follow (which was also known as the Shir Iran or Lion of Persia) was more commonly known as FV4030/2 Shir 1, and it basically was the front end of a Chieftain hull with a new engine compartment with a prominently sloped rear hull plate to accommodate the new power pack. The first three prototypes of the vehicle were ready and running during the early part of 1977, and in March of that year they were demonstrated to the Iranian Army The Shir 1 was vastly improved vehicle with up to 1200 bhp available from the new CV12 engine, a smoother ride (with the promise

In a revival of the old Chieftain Mk.5/2 concept, the Chieftain 800 and 900 were 1980-81 Royal Ordnance Leeds private venture proposals for Chieftain rebuilds. The proposal incorporated a new power train mating de-rated 800 and 900 horsepower versions of the CV12 with the Chieftain's TN12 transmission, Chobham armour, and in the case of the Chieftain 900, the Centaur fire control system. The upgrade programs were never proceeded with, and could have been offered to the export market as easily as to the British Army. The Chieftain 800 and 900 demonstrators were exhibited in 1981 and 1982 at the British Army Equipment Exhibition with mocked up armour. The project was abandoned by 1986 [Phil Richards].

A little-known member of the Challenger family that goes back to the very early days of MBT80 project: this is a test rig hull similar to the final design that was adopted for the Shir 2. The turret carries weights to simulate part of the 70 ton weight anticipated for MBT80 [BJ].

94 KC 36: a Challenger 1 Mk1 development vehicle, distinguished by the absence of a TOGS barbette on what is technically a production vehicle. This vehicle is maintained in full running order by the Imperial War Museum site at Duxford [Robert Griffin].

of hydropneumatic suspension developed by Dunlop and later Horstmann on production tanks) and the Fire Control System specified was to be the new Improved Fire control System that was just being introduced into the British Chieftain fleet, an excellent system for that time. The Shir 1 would however suffer production delays. There were problems with the new TN37 gearbox and these delayed both the Shir 1 and the definitive Iranian MBT which was to be the FV4030/3 Shir 2 (scheduled to be the first production Chobham armoured tank).

The MBT-80; the Tank that Never Was

Meanwhile the UK itself desperately needed a new tank to be developed after the collapse of Anglo-German project for an FMBT, and this became General Staff Requirement (GSR) 3572 *Replacement of Chieftain Tank*. Unusually, the decision to replace the Chieftain was given a vast amount of publicity, with the Daily Telegraph running headlines that intimated that millions of pounds were being spent to develop the new MBT

The first production Challenger 1 Mk1 on its handing over to the British Army at ROF Leeds. Although it has the TOGS barbette fitted, the actual units were not yet installed at this time [ROF].

A rear view of the first Challenger 1 Mk1 off the production line, ready to be issued to the British Army, notice the spare track links fitted the rear hull and how it slopes, in a great departure to previous UK tank hull design [ROF].

for the UK (given that the parliamentary enquiry into the L60 and its problems occurred that same year, 1978, it is easy to see why this may have been a public relations exercise by the MoD). The new tank project was officially designated MBT80, which loosely referred to the decade in which the new MBT would enter service.

No sooner was GSR3572 issued than it was surrounded by media hype, which quickly extended to the production of an A4 size brochure with a speculative drawing of the new tank on the cover, which looked suspiciously like Shir 2; this brochure was given to virtually every AFV crewman in BAOR. As an exercise in propaganda it may have served a certain purpose, but the MBT80 project had little substance as of yet, and remained years away from producing a definitive MBT. The newspapers were having a field day over the MBT80 announcement. But the problem was that it was just that, there were no finalised drawings, no officially confirmed ideas, no prototypes... just an announcement.

The MBT80 project set out to establish the basic design parameters and as usual various committees were formed and ideas investigated. One immediate matter to settle was the choice of main armament: the current 120mm L11A5, the experimental 110mm rifled gun or a new 120mm rifled gun. Power could be provided by either the Rolls Royce CV12, CV16 or a gas turbine. One novel feature was that part of the general design parameter was a combat weight of 70 tons; but part of the Chobham armour protection was to be removable; this was to enable MBT80 to stay within the current Military load classification of 60 tonnes. A basic weight of 50-60 tons was far more desirable for the design, however it was feared that the current and projected Soviet MBTs armed with the 125mm gun or its' derivatives could possibly already have the capacity to penetrate the MBT80 if it were left as a 60 ton specification.

To fit the armour needed to defeat the threat of the new Soviet125mm gun and new ATGMs the MBT80 would need to be heavy and wide, hence the decision to have removable armour arrays. The choice of weapon decided on was a development of the current UK120mm L11A5, designated the M13-A, it was a traditional rifled weapon but it introduced the split vertical sliding breech block that was to be a feature of the L30 gun fitted later to Challenger 2. The M13A gun was the piece fielded by the British in the trials determining the weapon to arm the new American M1 MBT, where it was in competition with the FRG's Rheinmetal 120mm smooth bore, and according to surviving records from the trial at the end there was very little to choose between the two weapons. The FRG's gun won in the end because the Americans preferred its simple one piece ammunition to the UK 3-part bagged charge ammunition.

The MBT-80 was to have a panoramic thermal image sight with the sight head located on the left of the turret (with independent traverse of 180 degrees controlled by the commander)

A good head-on view of the first Challenger 1 Mk 1 built for the British Army on its press day being officially handed over [ROF].

along with a laser stabilised panoramic sight for the gunner. This sight was built and tested and according to one of the Barr and Stroud engineers who worked on the development it was the finest set of optics he had ever worked with. An example of this system (known by the acronyms for the 2 sighting systems, PAN-O and PANTILI) survives today in the Aldershot military museum. The provision of this equipment in an advanced fire control system offered an effective hunter-killer capability in all light and weather conditions, and would have given the MBT80 the ability to detect and destroy enemy targets at the extreme limits of vehicle range.

The time scale for MBT80 availability was projected with the first prototype P1 to be ready by October 1983 and the remaining eight by 1985 (with a full training package to be available

A Challenger 1 camouflaged with freshly cut branches. Although this is forbidden today, during the major cold war exercises it was common practice, although frowned upon on in later years [DB].

A privately-owned Challenger doing its turn in the arena at a military show. If this was an actual in-service vehicle the loader would soon be in big trouble for his dress, and for sitting on top of the turret while the vehicle is in motion. The rule is to keep one's head and shoulders only above the roof level! [RG].

A Challenger on exercise in Germany, see how the crew have used standard army issue black masking tape to secure the plastic muzzle cap onto the end of the barrel of the 120mm L11A5 gun. In practise this could be left on, and blown away with the first round, but the caps never stayed on very well, hence the tape. The crewman is wearing the standard issue Army green denims with an assault vest and belt order and is carrying the standard SA80 series rifle [PH].

by 1987). Because the MBT80 specification was envisaged to incorporate new technology, it was decided to construct two Mobile Test Rig vehicles (MTR) to prove the technology as it was devised, so it could be incorporated already proven into the MBT80 prototypes as they were built. This, it was hoped, would save in development costs. MTR 1 was built from Chieftain components while MTR 2 was constructed from FV4030, or Shir components.

Only MTR 2 survives at Bovington but I must stress here that it does not resemble the production standard envisioned for MBT 80, although myths abound that attribute it as an unfinished MBT 80 prototype. The overriding reason that what survives of the program at Bovington is not a prototype MBT 80 is that the finalised hull and turret drawings had not been released by the time the project was cancelled, and no prototypes were ever built. Due to the secrecy, confusion and hype that surrounded the simultaneously running FMBT/MBT 80 and Shir 2 programs, a certain mythology has sprung up around the MBT 80, due in large measure to the media hype that surrounded it in the late 1970's and to a lack of information on the programs since. The truth about the end of the MBT 80 and the birth of the Challenger 1 is still surrounded with controversies to this day, and I hope that the facts we do know set out an adequate picture.

The Challenger is Born

In the spring of 1979 the Shah of Iran was deposed by the Ayatollah Khomeini and had fled the country, with dire consequences for the Shir program. One of the first acts of the new Iranian government was to cancel all the FV4030 contracts, and this put a large stress on the British tank industry, where ROF Leeds was already completing the Shir 1 orders for Iran. This left Royal Ordnance, Vickers and other contractors with completed

If you are going to get bogged in, then do it well! This crew, not content with simply getting stuck, have managed to also get their Challenger sunk at an angle. It is nothing that the REME recovery crew will not be able to cope with, but it will cost the tank crew a good crate of beer later [RG].

A misty morning on Hohne ranges in Germany. Range periods always involved early starts, to ensure that as soon as the range was open and clear, rounds could be fired to maximise the training time. This range may be already open as the crew can be seen wearing hearing protection, which is mandatory for firing drills [DB].

Although this looks like a conventional Challenger 1 there is a subtle and very important difference: it is fitted with the L30 120mm gun of the Challenger 2. Prior to 1990 the L30 upgrade was going to be part of the Royal Armoured Corps' policy of having two types of MBT in service, although it was abandoned in the wake of the breakup of the U.S.S.R. and the dissolution of the Warsaw Pact [ROF].

"Oh whoops" could be the catch phrase for this crew and their well bogged-in Challenger 1. This will be a job for the Challenger recovery vehicle, but a lot of wet mud will have to be dug out by hand first, not a pleasant task [SC].

Shir 1 tanks on hand and advanced development programs and large investments made for the Shir 2 program. Britain could not absorb the Shir 1 into its own defence procurement because it was not suitable to the future requirement for a new British MBT, but Royal Ordnance had to dispose of them in an cost-effective manner or a number of companies faced financial crisis: they were only saved in this by the purchase by the Royal Jordanian Army of the 274 existing Shir 1 tanks (which was renamed Khalid), and deliveries commenced in 1981.

Meanwhile, development work on the MBT 80 project carried on, although there were calls from some quarters to abandon it and purchase either Leopard 2 or M1 off the shelf. The purchase of a foreign tank gained little support and was not followed through for two reasons; one obvious reason was the preservation of the defence industry and the other (more often quoted) was that because Leopard and Abrams were either entering service or would be very soon, their technology would be dated by the time of the in-service date of MBT 80 (projected for 1989). The embarrassing part about this date was it would mean that although the UK had invented Chobham armour, over twenty years would have passed before the UK fielded an MBT equipped with it. By 1979 the USA would already have M1 tanks in service fitted with Chobham armour due to Britain's sharing of this technology with its main NATO allies in the early 1970's.

The problem was that while the MBT 80 project was taken seriously it was quickly recognised that the estimated date for MBT 80 deployment was slipping past 1989 due to anticipated delays involving some of the systems and funding for the new tank, and it was feared that the UK might not get its new tank until sometime in the 90's. The official position within the MoD was that MBT 80 would enter service on time, but this was doubted in some political quarters as other pressures dangerous to the MBT 80 program emerged. The deployment of large quantities of new tanks such as T-64 and T-72 from the Soviet Union, and the spiralling cost estimated for each unit of the MBT80 were also served to put the MBT 80 program's future in doubt. Weighed against the substantial investment already made in the Shir 2, the decision was taken in September 1979 to procure a quantity of the FV4030/3 (modified to British requirements but embodying a minimum of design changes to enable rapid procurement) to supplement the Chieftain fleet. This in turn enabled the government to quietly sideline the MBT 80 project, which carried on briefly with several major parameters yet to be firmly established (such as the type of engine to be procured) prior to eventual cancellation.

The decision to hastily procure the FV4030/3-based MBT without any major modifications, would prove to be short-sighted in later years. Although the MBT 80 had been cancelled, various small projects derived from it ran alongside the Challenger program for a few years, mainly to prove various components for future use. A wiser program would have seen the FV4030 hull and automotive systems allied to a new turret on an MBT protected with Chobham armour, but in 1979/80 funding and the political patience necessary were lacking.

Although the media trumpeted dramatic headlines announcing the government's commitment to the new MBT and applauding the billions of pounds spent on the new tank, those on the inside knew that the FV4030/3 based tank was not quite the answer and that it would not be acquired in large numbers. Only 243 tanks were initially to be purchased; only sufficient to equip four armoured regiments, and the remainder would soldier on with Chieftain, which was to be upgraded, with a Mk.12 version specified and designated (which in turn would never be proceeded with). This was not an ideal state of affairs because it meant that logistics in the field would have to deal with two types of MBT.

There were also major design problems with the Shir 2 to be dealt with as well, especially with the TN37 gearbox. There was a certain ignominy in the trials of the new MBT, when on trials at Chertsey, not being allowed to drive across a road bridge over the main civilian road under its own power due to the unreliability of the transmission (it was towed across by a venerable Conqueror tank, due to fears that if the gearbox failed it could potentially send the tank crashing down onto the road). In October 1982 an exercise took place on Salisbury Plain which involved 4 prototype FV4030/4 Challenger MBTs (as the vehicle would be named), which proved that the vehicle was acceptable to the British Army. On the 14th December 1982 the Challenger was accepted by the general staff, but with several issues to be resolved. The choice of Challenger for a name has caused many stories to be told, for example, the question of why a name from a rather unsuccessful WW2 cruiser tank was used again. The truth is that once it was known that the new MBT was going to be issued it was decided to hold a competition within the RAC to name the new tank, and the winner was S/Sgt Larry Murray of the 4th/7th Royal Dragoon Guards who choose "Challenger".

The Challenger Described

The layout Challenger offered was nothing controversial or new. It was basically a Chobham armoured, re-engined, incrementally modernised and improved version of the Chieftain that preceded it. It had a four man crew and adopted the supine driving position pioneered on Chieftain, it had the same 120 mm L11A5 series gun and a similar fire control system (based on the Chieftain's IFCS) which was to be judged harshly versus its NATO contemporaries. The obvious improvements that the Challenger offered over the Chieftain were the replacement of the L60 by the Rolls Royce (later Perkins) CV12 diesel, the new hydro-pneumatic suspension; and most importantly the new Chobham armour protection.

The Challenger's armour was of great interest to Britain's allies and potential foes. When the Challenger was first issued to units their security arrangements had to be raised significantly. Whereas by 1980 broken down Chieftain tanks would often be left on the vehicle park over a weekend if they required a power-pack change, due to the secrecy surrounding Chobham armour this was no longer an option, so broken down tanks had to moved back inside their hangars, all hangars were securely locked and the tank park itself was enclosed by high wire fences and guarded (just as had happened when the Chieftain first came into service). Military enthusiasts were desperate to discover the composition of Chobham armour, not for any malicious purpose but simply to be the first to write that they had found the secret. These stupidities extend to some being caught at military expositions featuring Challenger prototypes, scraping the armour for samples. This was all a waste of time because the actual armour array was hidden by external cosmetic plates.

The Challenger hull layout resembled the Chieftain's, with the driver sitting centrally with a lift and swing hatch providing access, and in certain positions of the turret's traverse he could gain access through the fighting compartment. He had an accelerator pedal and brake pedal in the normal locations and a foot rest for

A troop of Challengers from The Queen's Royal Lancers, stopped on exercise somewhere in Germany in the mid 1990s. The three crewmen on the back decks all have a tin of rations in their hand, so this is possibly a late breakfast. These tins will have been heated in the Boiling Vessel and may have been in there for some time, keeping warm until an opportunity arose to stop and eat [DB].

Completely unacceptable today from an environmental perspective, but at one time this was the normal thing to do on exercise, to create a mobile armoured tree!! The problem with using live camouflage is it tends to die quickly and will thus look out of place in a sea of green. Although the bushes on the tank are different from the tree behind it is obvious that from a distance how the two will blend in and help hide the tank [RG].

his left foot, which in Chieftain was used to change gears, but the Challenger's automatic gearbox the driver used a pre-selector gear shift, giving a range of four forward and three reverse gears, all of which eased driving tremendously. On his left and right side of the driver's position were boxes containing the vehicle batteries, and above the left set of batteries was a combined stowage box, which held general stowage in one part and three 120 mm projectiles alongside. Located on the right of the drivers compartment on the hull floor, were controls for the auxiliary generator and emergency gears, these were easily located with the drivers hatch open but were difficult to reach once driving closed down. On either side of the driver's seat were the steering levers which were adjustable to allow opened up driving or closed down. The driver was provided with a No 36 Mk1 day periscope, cleaned by a washer and wiper system operated by the driver. For night driving the driver changed the periscope for an L41A1 Image Intensified (II) sight, a simple operation for the driver to carry out. Communication with the rest of the crew came from the intercommunication system integrated into the Clansman radio system fitted as standard in all Challenger 1 MBT's.

The loader/operator, gunner and the tank commander were stationed in the turret. The loader's position was located on the left of the turret, the gunner on the right and behind and above him sat the vehicle commander. The turret layout followed that of the Chieftain, which would be criticised for its poor ergonomics in the later part of the 1980s. The focal point of the fighting compartment was the breech and recoil system of the 120mm L11A5 gun; the same weapon as mounted in later marks of the Chieftain. The breech was opened initially by the loader, using the breech mechanism lever located on the right of the gun; at all other times the breech opened automatically during the gun's recoil run out. The gun used split ammunition exactly like that used in the Chieftain. On firing the gun recoiled approximately 35cm under control of two hydropneumatic buffers.

The only part of the ammunition remaining after firing the 120 mm gun was the brass case of the vent tube. The bag charge was consumed totally completely during the course of firing, which was one reason the UK carried the L11 system over into the Challenger. Ammunition stowage followed the pattern employed in the Chieftain, with water-jacketed charge bins provided for the propellant charges, all stowed below the level of the turret ring. The later charge bins fitted to Challenger 1 Mk.3 dispensed with water jackets as used on the Chieftain and used steel armoured bins instead, these were standard for the vehicles used in Operation Granby and were carried onto most of the BAOR Challenger fleet thereafter. Projectile stowage followed the late Chieftain pattern, with APFSDS munitions as the initial weapon of choice against enemy MBT's. The Challenger 1 was also provided with HESH projectiles and was often stocked with APDS ammunition, which had been procured in large quanitity during the initial Chieftain program and was used for practise or specified for lesser armoured targets like older enemy MBT's. As well as looking after the L11A5 the loader was also responsible for loading and clearing stoppages in the 7.62 GPMG co-axial machine gun, and if that is not enough he also assisted the vehicle commander in taking and decoding radio messages and encoding outgoing messages. The loader was also is responsible for maintaining the radios and ensuring that they are always on the correct frequency. To ensure that the loader did not become bored he was also responsible for ensuring that a constant stream of tea or coffee was available and cooked the crew's food in the boiling vessel (an electric element in an insulated box, with the capability to boil water for drinks or cook food for the crew whenever needed).

A rarely seen shot of a Challengers rear turret bin with the lid open. It is strange why the designers chose to have the lid open to the rear instead of going forward to ease access for the crew from the engine deck. The drum to the right contained a two core cable known as D10, which allowed the crew to operate remote communications up to 100 meters away from the tank [RG].

The job of operator/loader was usually the last of the progression of crew positions before a tank soldier took the big step into the commanders' seat. The loader accessed the turret through a hatch with two spring loaded doors in the turret roof. The only view he had once fully closed down was via his Periscope AFV No 30 Mk1, located in the forward part of the turret roof. One advantage the loader had was the small area he could move around inside the turret which compensated for the small and uncomfortable seat he sat on, a comfort the other three Challenger crew members did not have.

The most confined crew position in the Challenger was that of the gunner, located on the right of the turret, below the commander. He had the massive 120mm breech to his left, the main gun sight to his front, and on his right the main gun control panel and TOGS viewer. The main sight used by the gunner for shooting and observation was the Tank Laser Sight (or more properly the Sight, Periscopic, Laser Rangefinder, AFV No.10 Mk.1), an advanced version of the same laser sight found on the Chieftain. The TLS was used to lay onto the target and then (by means of the laser) determine the range. The range was displayed both in the gunners and commanders sights, but most importantly the range was part of the data fed into the computerised fire control system, designed to give a good chance of first time hits. The system also receives information on the speed of the tank (if moving) speed of target, barrel wear and attitude of the tank, all this fed into the system and the result is projected into the gunners sight by the means of an ellipse which will appear around the target and the gun will be driven to the correct elevation and line. After firing corrections are applied via the computer/manually, and although the system was dated by the time of Challengers introduction it still could give a good account of itself if allowed to.

The gunner had controls to allow him to traverse the turret and elevate or depress the gun under power or by hand. The power control was located to the gunner's right; it consisted of a handle with a small thumb controller located on the top; controlling both traverse and elevation. Located nearby was a hand traverse handle for use as a back-up if the power equipment had been damaged, or for fine laying. Immediately below the laser sight was the elevating hand wheel, and to its right was a fixed firing handle incorporating a selector switch to allow the gunner to select either the main or co-axial armament. A second control used by the gunner fed the range to the computer and generated the automatic laying of the gun. The gunner could also either fire the co-axial machinegun electrically by use of the firing switches, or by use of a foot pedal located near his right foot.

The Challenger's Tank Laser Sight No.10 Mk.1 used a binocular arrangement for the gunner, and had a magnification of X1 for general viewing and X10 for engaging targets. When the gunner established laser range using the sight it was displayed in his left eyepiece, with the right eyepiece being used for viewing (this range readout was also visible to the commander in his range readout unit). The sight also contained part of the muzzle reference system, or MRS. This very simple system allowed the gunner to maintain gun/sight relationship at all times. Prior to this the gun would be bore-sighted first thing in the day, which would enable all the sights to be aligned correctly with the bore of the gun, but if the barrel should strike a tree or similar object this relationship could well be lost and would affect accuracy. It was decided some method of checking alignment while under armour was required, and the solution was simplicity itself. The system consist of three main components, secured to the muzzle end of the barrel was a rubber shroud, under which was mounted a highly polished metal mirror on an adjustable bed. The mirror

This Challenger 1 belonged to the Demonstration Squadron of the Land Warfare Centre at Warminster, and the unit pictured is D Squadron, The Royal Dragoon Guards. They act as enemy forces OPFOR on Salisbury Plain, as well as being used for almost any armour-based activity including the training of young troop leaders. Their Challengers were painted in similar colours to the BATUS vehicles to distinguish them from other users during exercises [RG].

reflected light projected into it from the light source located under the gunner's hood, alongside the main gunner's sight. The light source was a 24 volt halogen lamp directing light into a large diameter fibre optic cable covered by a red filter, shaped to conform to the shape of the 120 mm bore sight marks found in the gunner's sight graticule. The final components were the depression indicators. These were mounted on the gun cradle and on a static bracket to the left of the gunner's sight, providing a simple method of levelling the gun to any set elevation to use the MRS.

To use the system the gunner simply elevated or depressed the gun until the indicators were aligned, then used the bore sight check lever located on the laser sight, and moved it to the check position, allowing him to see the reflected aiming mark from the mirror. He then switched on the light source and could

This shows the red painted recessed fire extinguisher handle of the Challenger's fire suppression system; there were two of these, on the left and right sides at the glacis joint. Connected to two internal fire extinguishers that were linked to the engine compartment, in the event of a fire these could be set off either by a crew man on the ground of from similar handles inside the tank. The picture of the Grim Reaper seen here on *Dark Angel* was deemed unsuitable soon after this shot was taken and had to be removed. Another tank from the same squadron was named *Devil May Care* [RG].

A general view of the Royal Dragoon Guards tank park at Warminster, scenes like these are played out in most armoured units with only the location changing [RG].

see an image of the 120 mm bore sight mark in his sight, then used the graticule adjusters on the sight to adjust the graticule until its MBS (muzzle bore sight) mark was in line with the projected image. Once this had been done he then reset the gun controls and was ready to fire.

The gunner was also provided with a secondary sight in case of damage to the laser sight, the sight unit AV No.87 Mk.1. The sight was located on the turret roof and was used by the gunner after releasing an armoured cover that allowed the sight head to protrude from the turret roof which happened as the gunner pulled down on the main sight body. The secondary sight had a similar graticule pattern as on the laser sight's viewer but had no range finding capability and was a basic gunnery sight only. Challenger 1 had one big advantage over all but the last mark of Chieftain in foul weather or night fighting. Challenger t was fitted with the Thermal Observation and Gunnery System, better known as TOGS, which although developed primarily for night fighting was equally at home in daylight. TOGS represented a massive leap forward compared to previous night viewing systems, such as the infra-red and image-intensified viewing devices fitted to the Chieftain series.

To explain the workings of thermal imaging (or TI) is way beyond the scope of this article so only a very simple explanation follows. TI works on the principal that all objects radiate

A Challenger 1 crossing a bridge built by 35 Engineer Regiment, Royal Engineers. Notice the speed limit signs and also how close the bridge sides are to the tank [DB].

In the mid-1990s the rear fuel drums introduced for Operation Granby quickly became the target for resourceful crews looking for handy stowage spots. Empty drums became a perfect way to increase the onboard luggage space, a tradition continued on the Challenger 2 [RG].

A rather unusual view of a Challenger: this is one of the few that remained in UK hands after its withdrawal from service in 2000. It is used as a recovery training aid, and has been pulled into this position by two Challenger recovery vehicles [DB].

Another sad end for a Challenger tank (although it could be said that it is still serving its country as a training aid). If this was a real situation, either the recovery team or the tank crew (or both) would have to get very wet; especially in fitting the tow bars or cables to enable the drowned tank to be recovered. Very few Challenger 1s escaped sale to the Jordanian Armed Forces after 2000 [RG].

some heat, which can be viewed with the right equipment and for Challenger that was the Thermal Image Sensor Head or TISH; located on the right side of Challengers turret and slaved to the main armament. Any thermal radiation picked up by the sensor were turned into electrical signals, processed into a viewable image available to both gunner and commander; through the Symbology Processing Unit (SPU) known to turret crews though as the *spew*! This unit also took all the data from the TISH sensors and used it to produce a thermal gun sight image with ballistic graticules in both the gunners and commanders TOGS viewing screens. The TOGS system proved itself as a great battle field asset (as was shown in Operation Granby) and was carried over in modernised form to the Challenger 2. The TOGS system was retrofitted to the first batches of Challenger 1 MBTs delivered (which were designated Mk.1s until fitted with the TOGS system) because the tanks were ready before the thermal sighting systems could be delivered. Once fitted with the TOGS system these vehicles were re-designated Mk.2s.

In the final crew position, the vehicle commander sat to the right rear of the turret behind the gunner, under his cupola with his own sighting equipment and power controls for the main armament. The cupola No.32 had nine fixed periscopes giving the commander a 360 degrees view, while the 360 degree traversable top plate of the cupola incorporated a spotlight, a second 7.62mm GPMG with its ammunition and the commanders main vision device, the Sight, AFV No.37 Mk.6. This viewing device incorporated a X10 sight for engagement and X1 for general battlefield surveillance.

One of the disappointments of the Challenger was that this sight was fitted for the British version of the FV4030/4 instead of the Condor day/night sight fitted to the Jordanian Khalid. A "swap" sight was produced for the Challenger commander's vision at night which looked very much like the No.37 sight, but as its name suggested it was a case of removing the day sight and fitting in its place the Sight, Image Intensified L5A1. The process of switching sights was never well- liked and contributed to the image-intensified commanders sights being ignored where possible, and in the end very few image intensified sights were ever fitted to the Challenger 1 in peacetime. The commander could take over the whole engagement sequence if he wished, although it increased his workload it was an essential ability if the gunner became injured. The commander's seat was on a hydraulic lift that enabled him to pump the seat up so his shoulders were only just flush with the rim of the cupola when operating opened up (which is the correct position for tank commanding!). It also made commanding the tank more comfortable for him on long exercises.

A significant improvement over Chieftain introduced on the Challenger was higher speed and mobility, thanks largely to the automotive system chosen for Challenger and its suspension system. Instead of the woeful British Leyland L60, the Challenger 1 was fitted with a Perkins diesel CV12 engine (originally produced by Rolls Royce), this was combined with the cooling system and the David Brown TN37 transmission into a power pack, which could be removed as one unit for repair or replacement. The CV12 gave Challenger 1200 bhp and a power to weight ratio of 20 bhp/tonne. The pack had generators to supply electrical power to the vehicle, but to use those the engine had to be kept at constant revolutions so the turret electrical systems also drew power from a generating unit engine (commonly called the "genny") derived from that fitted to Chieftain, the Generating Engine H30 No.4 Mk.18. With the "genny" running, the vehicle was provided with full electric power to run all systems, especially useful at night when TOGS may be running (and of course the boiling vessel). The replacement of the Chieftain's old Horstmann suspension by the use of Dunlop hydrogas units gave a far better ride and the suspension weighed only 75% of the total weight of the older units. While the Leopard 2 and M1 series boasted higher speeds than the Challenger's specified 56 km/h, the Challenger was easily able to exceed its theoretical top speed and the type's excellent battlefield mobility carried on to its eventual successor.

A practice Armour Piercing Discarding Sabot Tracer round. This round, which is painted Saxe Coburg Blue, was used as the training round by Chieftain and Challenger gunners, although both tanks could fire the later fin-stabilised rounds. The round is weighted to represent a service APDS round, and behind it can be seen the full bag charges for the round [RG].

The practice bag charge for use for gunnery training drills with APDS/T rounds: the outer casing is made of Nitrocellulose craft material, which is waterproof. Large quantities of APDS/T munitions were procured in the 1960s and 1970s, and these persisted as part of the ammunition issue for the Challenger as well, even though more effective APFSDS/T was available from the early 1980s onwards. Early L11 series bagged charge bags at the beginning of the Chieftain program were all made of either silk or calico, and if the bag charge container leaked it made the charge unusable. This problem was rectified long before Challenger entered service. The blue end incorporates the lifting handle and also the igniter charge of G20 gunpowder, which ignites the main propellant charge [RG].

A charge bin removed from the tank. This is the loaders ready bin from which he will take charges first, and it can hold either 3 APFSDS charges or 6 HESH charges. It can also hold SH/PRAC/Smoke charges or any combination of them. The clear plastic closure lid allows the loader to see at a glance what charge is inside [RG].

Although obviously posed, this shot does give an idea of the loading drills and the size of the HESH round. This is a service round, denoted by the yellow/black markings, practise rounds will be painted Saxe Cobourg Blue, similar to the DS/T rounds in the forward rack. Behind the stowed HESH rounds can be seen two silver coloured boxes, these contain the vent tubes and are only removed when they are going to be loaded into a magazine [British Army].

The commander's cupola and associated equipment, the General Purpose Machinegun (or GPMG) can be fired from under cover by the commander, using his x1 window in his commanders sight. The belt fitted in the picture is obviously made up of empty cases for safety reasons. Normally three boxes of 200 rounds of 7.62mm NATO would be stored in the mounting on the left of the cupola, under a plastic cover to keep rain and dirt out. In front of that is the cupola spotlight, and immediately below it can be seen the head of the gunners secondary sight [RG].

A view from the commander's cupola looking forward, showing the rear of the cast commander's sight hood [RG].

The ammunition for the cupola 7.62mm NATO GPMG. The GPMG could be fired closed down and the ammunition was kept clean and dry by the plastic cover shown [RG].

The inside of the commander's cupola hatch, showing the padding, latch and grab handle [RG].

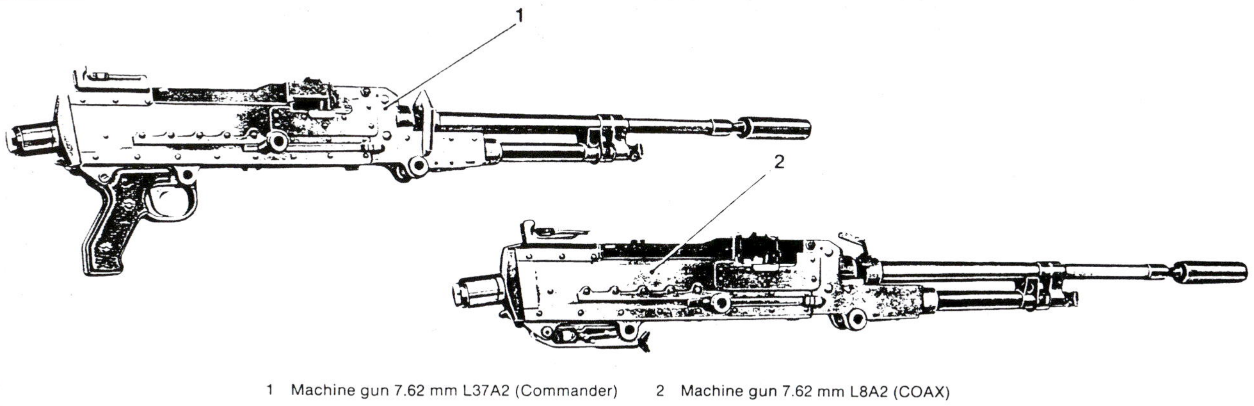

The secondary armament carried by the Challenger 1 MBT throughout its service comprised two variants of the MAG-58 derived General Purpose Machine Gun, or GPMG, a belt-fed 7.62mm weapon of Belgian origin which has long served the British Army. The L37A2 variant is mounted on the commander's cupola and can be used if necessary in the dismounted role by fitting a butt stock , while the L8A2 is a modified weapon optimised for use as a co-axial machine gun [Crown Copyright].

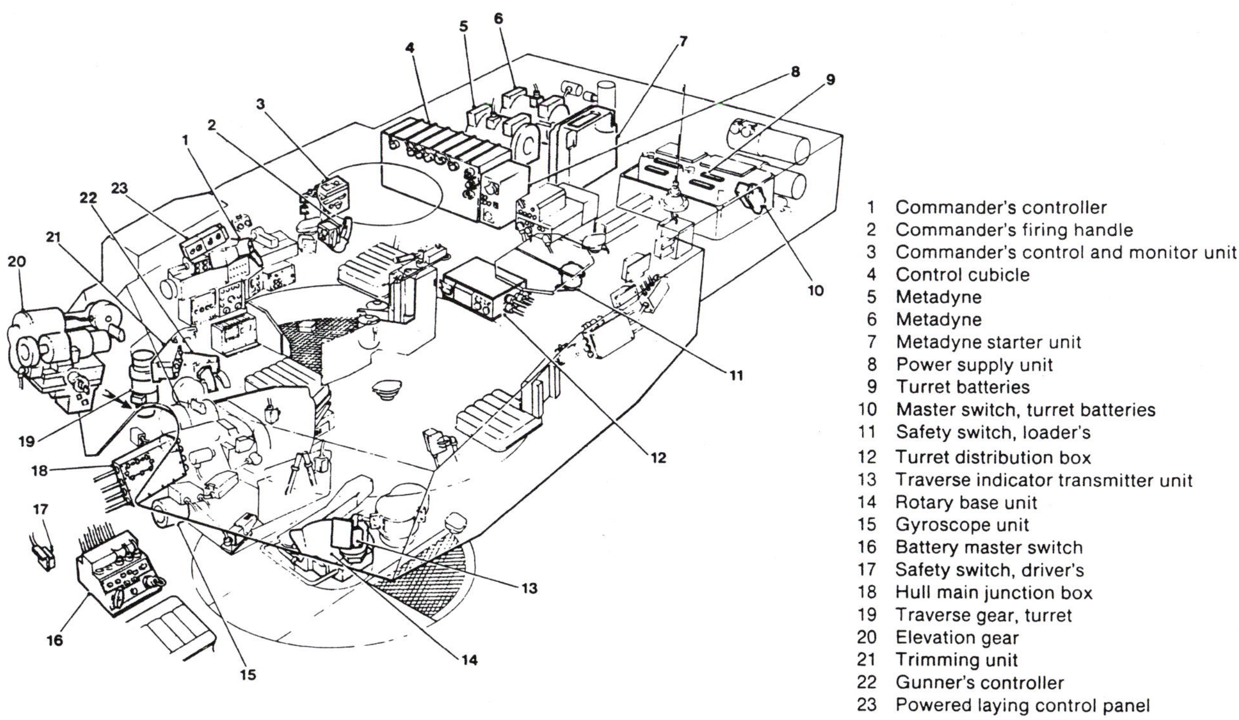

A simplified drawing showing the gun control equipment components and their locations around the Challenger 1's turret. The Challenger 1 Mk.2 was criticised for poor turret ergonomics in the wake of the Canadian Army Trophy competition of 1987, and the Mk.3's turret layout incorporated some changes [Crown Copyright].

The commanders cupola from the right front [RG].

Another view of the front side of the commander's cupola, the gunner's secondary sight periscope can be seen below the GPMG [RG].

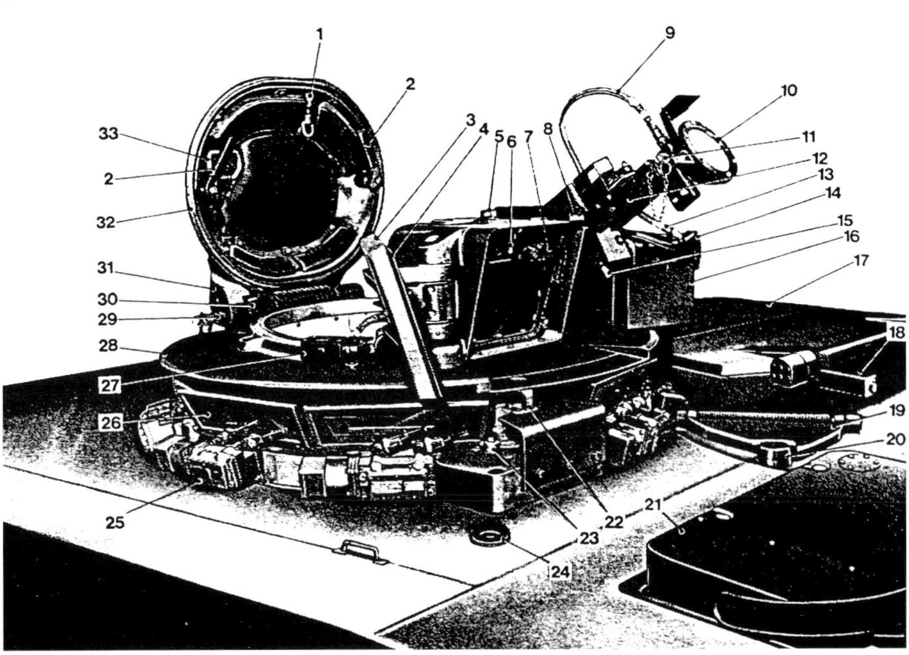

1 Bolt
2 Hatch cover locking handle
3 Sight hood door
4 Shield
5 Reservoir filler plug
6 Washer spray
7 Wiper unit
8 Door release
9 Spotlight cable
10 Spotlight
11 MG front mounting pin
12 MG cradle
13 MG solenoid cable
14 MG ammunition box cover
15 MG depression stop
16 MG ammunition box
17 Loader's hatch
18 Hatch catch
19 Auxiliary sight cover
20 Sight cover lock
21 Gunner's sight hood
22 Commander's sight washer jet
23 Periscope washer jet
24 Compensating link filler plug
25 Wiper motor
26 Periscope
27 Sight hood connector box
28 Cupola
29 Quick release pin
30 Hinge lug
31 Catch arm
32 Cupola hatch cover
33 Spade catch

A manual illustration showing the parts of the commander's cupola [Crown Copyright].

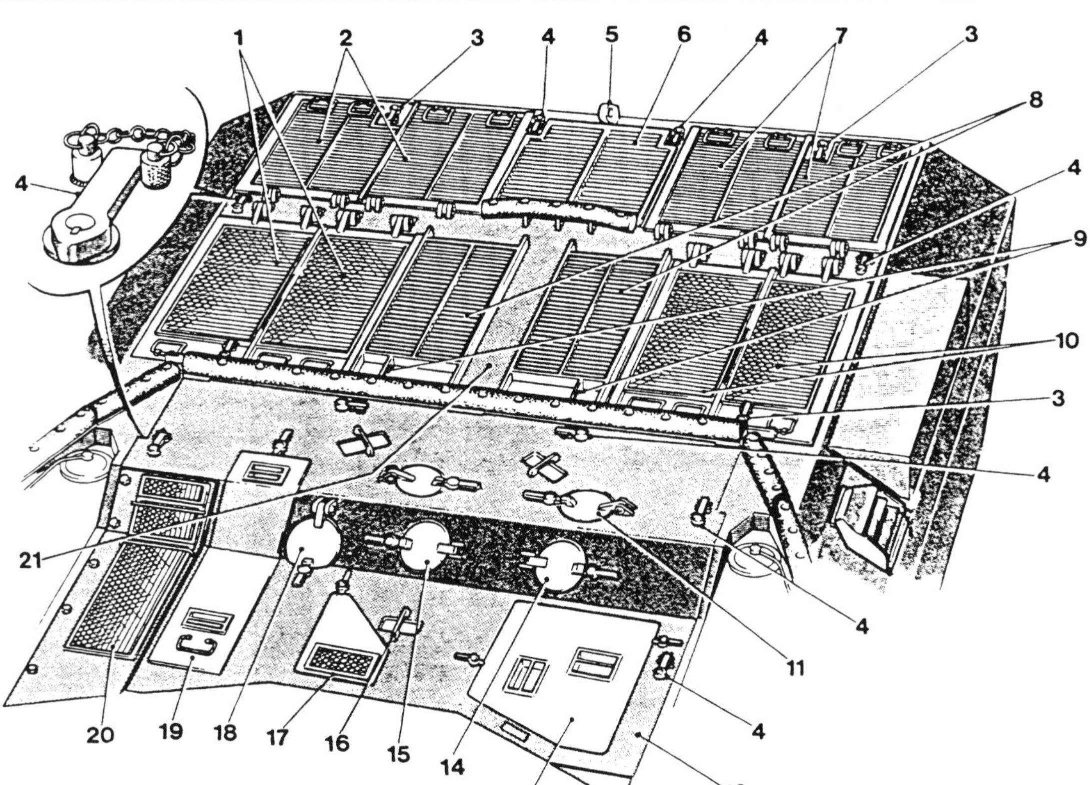

1 Radiator access louvres (hinged)
2 Rear compartment access louvres (hinged)
3 Louvre catches and securing pins
4 Deck securing catches and securing pins
5 Lifting lug (rear)
6 Air outlet louvres (fixed)
7 Rear compartment access louvres (hinged)
8 Air inlet louvres (fixed)
9 Lifting lugs (rear deck, front)
10 Radiator access louvres (hinged)
11 Coolant filler access
12 Front deck
13 GUE air cleaner access
14 GUE oil filler and dipstick access
15 Fuel solenoid override access
16 Lifting lugs
17 Ventilation air inlet
18 Main engine oil filler access
19 Main engine air cleaner access
20 Main engine air intake grille (hull breathing)
21 Rear deck

A manual diagram showing the engine deck layout and components. While the hatches incorporated in the engine deck allowed access to the air cleaners, engine, transmission and cooling system, the entire engine deck could also be removed as a single plate [Crown Copyright].

The empty engine bay of a Challenger hull in private ownership being restored. Very few Challengers have made it into private collections due to issues in removing the Chobham armour packs [RG].

A Perkins CV12/TN37 Powerpack seen outside of the tank. The CV12 generated 1200 BHP, and the Challenger 1 could easily exceed its theoretical top speed [RG].

The Perkins CV12/TN37 Powerpack could be changed more quickly than the Chieftain's L60/TN12 combination, but was a significantly larger power unit [RG].

The right hand rear edge of the engine deck houses the right hand fuel filling point with its protective cap and rail [RG].

The brackets for the pioneer tools on the right side hull pannier nex o the engine deck. The large access plate is to access the right fuel cell [RG].

The radiator access louver doors on the right hand side of the engine deck [RG].

The right hand side air inlet louvers on the engine deck [RG].

On the left hand side of the engine deck, the radiator access louver doors (left) and the left hand air inlet louver (right) [RG].

On each side of the engine deck behind the exhaust outlets a large stamped stowage bin was mounted [RG].

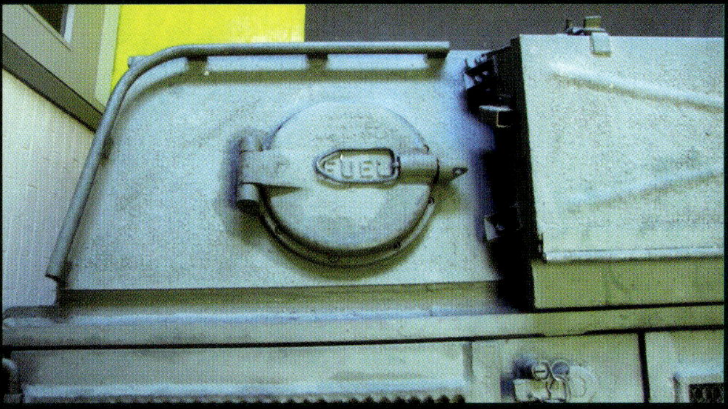

The left hand diesel fuel filler with its large cap aft of the stowage bin [RG].

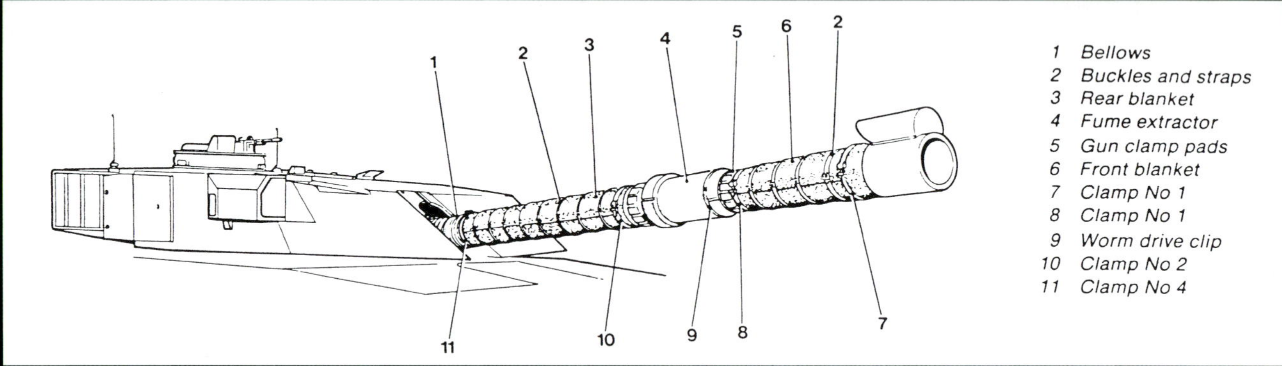

The L11A5 120mm gun barrel, like that of the L11 guns fitted to the Chieftain MBTs, was fitted with a thermal cover which prevented changes in temperature from affecting the gun's accuracy. This diagram shows how the two part cover was secured. A fume extractor was fitted mid way up the barrel [Crown Copyright].

A section through the breech end of a 12mm L11A5 barrel, you can see how the APFSDS round sits snuggly into the neck of the nitrocellulose bag charge. Notice the rifling does not start until just level with the fin round's top [RG].

An instructional cut away section of the L11A5 120mm gun breech. The brass vent tube can be seen attached to the firing needle, once ignited the flame will travel up the white painted portion visible and ignite the combustible bag charge, and all that will be left after firing will be the .625 vent tube. On top of the breech can be seen a set of obturators which seal of the breech during firing [RG].

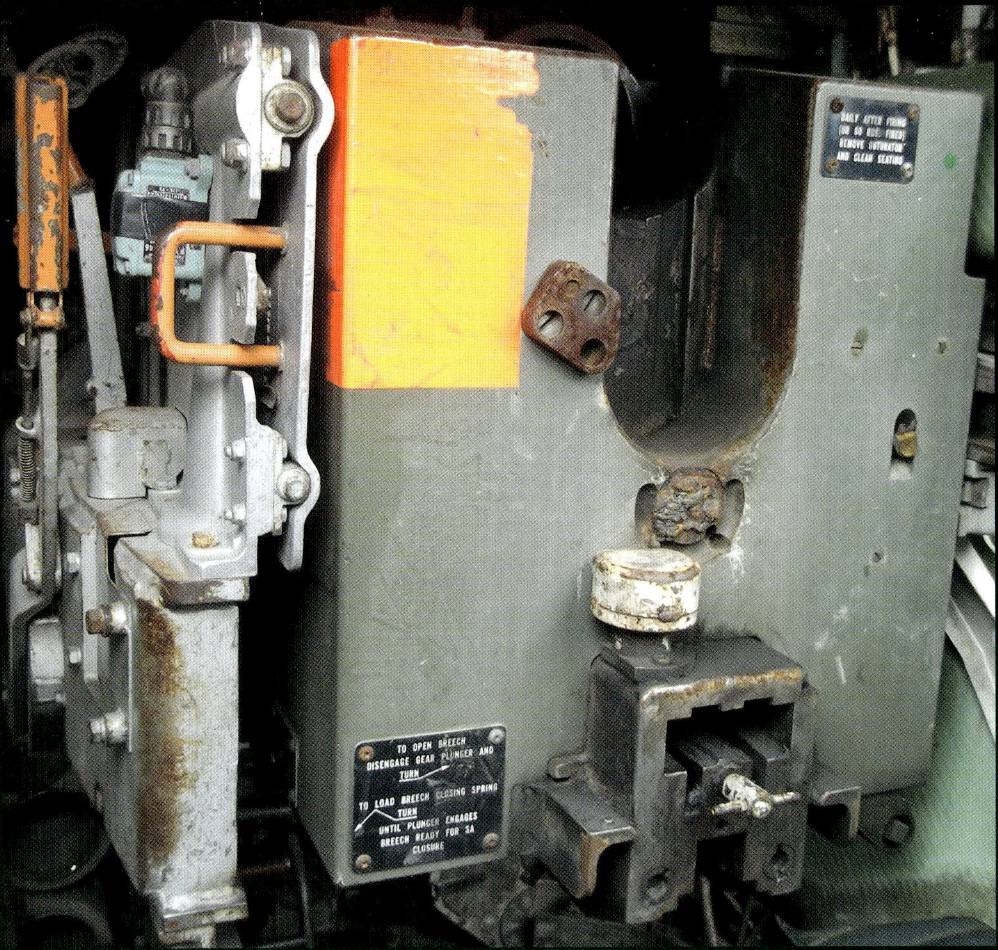

The rear view of the 120mm breech ring, showing from left to right, the large orange handle used to open the breech manually, the smaller orange handle used to pull the loaders safety shield to the rear, and the vent tube loader located at the rear, with the white topped plunger located on top of it. The magazine of vent tubes would fit below this and the loader would use the white plunger to load them. The tank pictured is decommissioned and has parts of the breech welded to comply with U.K. laws. Above the plunger is the seating for the firing needle assembly which would receive the vent tube once the breech closed [SC].

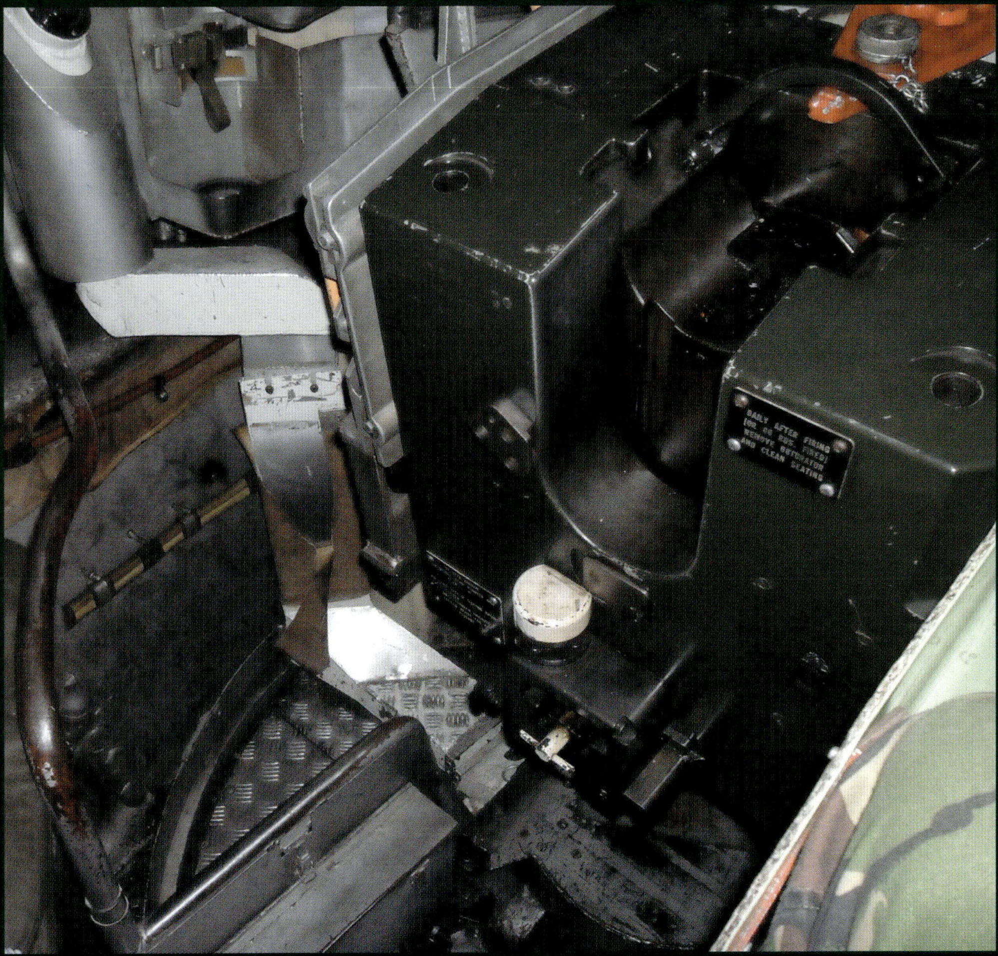

Looking down from the Commanders seat one can see clearly the vent tube loader, the lower white handle is the actual rammer that will force the vent tube into the breech block when it is aligned, the top handle is the plunger which is operated by the loader. Once the gun has run out, the rammer will be held to the rear by the plunger, this was introduced to prevent the tube being misfed, as initially the rammer would automatically load the next tube. It was found that the breech block was still bouncing and often caused alignment problems, so now the loader hits the plunger and also the rammer as part of the loading drill [RG].

The rubber bellows protected the base of the gun barrel from the elements where it emerged from the gun mantlet and prevented dust from entering into contact with its surfaces when the gun recoiled and ran out afterwards [RG].

The mantlet cover plate. Note how the barrel of the coaxial machine gun is situated above the main gun on its left left side, just as it had been on the Chieftain. The cover plate has a cut out for a lifting eye [RG].

The mantlet cover plate was secured by 5 bolts and elevated with the main armament [RG].

The gunner's main sight seen from the front with the armoured visor lowered [RG].

The gunner's main sight seen from the rear [RG].

The gunner's secondary sight head showing the protective flap and the retaining catch, which when undone allows the spring to open the cover, but which must closed again manually from outside the turret [RG].

The right front hull access plate [RG].

This shows the location and mounting for the splash plate, also note the thickness of the plate [RG].

Front left headlight, located directly behind is the vivid red fire extringusher that's has to be that colour to comply with EU law. A green plastic cover was available to cover it for exercises [RG].

Here can be seen the spring loaded arm for the drivers mirror, the tube behind the sidelight is for the camouflage poles [RG].

General view showing, nearest to camera, side light, camouflage pole bracket and headlight guard, notice also the anti slip finish [RG].

Front right bollard, this is retained by the pin and as can be seen the end of the pin drops down to retain the bollard. When being moved by tank transporters the bollards are removed and the tie down chains are fitted to the bracket [RG].

The bazzoka plate shoing the mounting bolts and also of interest to the modeller are the lifting handles which are made of flat metal and not round as sometimes depicted [RG].

The gun clamp in its stowed location, to release it the catch on the right is pushed forward and the clamp swing up [RG].

A top view looking at the rear jerry can stowage position, the pin and hinged plate were used to retain the water cans [RG].

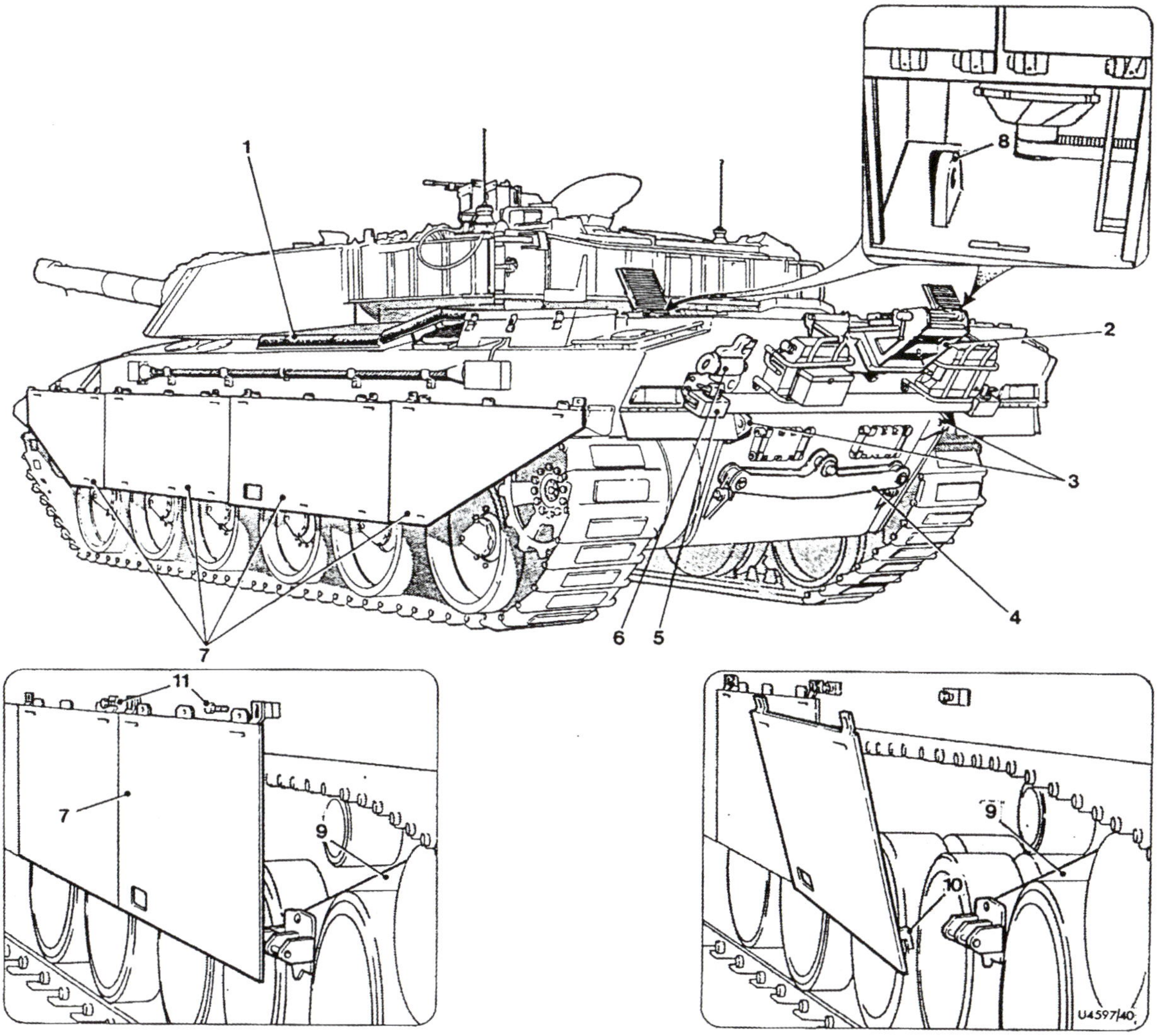

1 *Gun depression rail*
2 *Gun crutch*
3 *Lashing eyes*
4 *Towing beam*
5 *"A" frame leg*
6 *Towing swivel*
7 *Skirting plates*
8 *Lifting eyes*
9 *Outrigger brackets*
10 *Tongue*
11 *Securing bolts*

A manual illustration showing the rear of the Challenger 1 and the method used to attach the bazooka plates [Crown Copyright].

In the foreground is the plastic cover of the basket, while below it can be seen the top of the side mounted exhausts, also visible are the parts of the resiliant rubber rail and at the end of the rail can be seen one of the circular access panels to the fuel panniers [RG].

A rear view of a Challenger 1 Mk.3 with the L11A5 gun secured in the gun clamp. The brackets for carrying additional fuel drums can be seen, these were fitted to tanks that served in Operation Granby as well as to many of the other Challenger 1s in the British tank fleet during base rebuilds or at unit level in the 1990s [RG].

The fittings found on the right side of the engine deck aft of the exhaust outlet [RG].

The diesel fuel filler on the right rear corner of the engine deck [RG].

Looking from the loaders side towards the gunner's and commander's stations, the first impression is one of a jumbled mess. It must be borne in mind that this is how the vehicle looked the day it was taken out of service, but various controls are easily spotted, among them the commander's traverse/elevation controller, which is just left of the centre and it is the grey pistol shaped component. To the right of the commander's seat is the commander's box which is used to input data to the IFCS onboard computer, such as range corrections, and it also will show fault locations [RG].

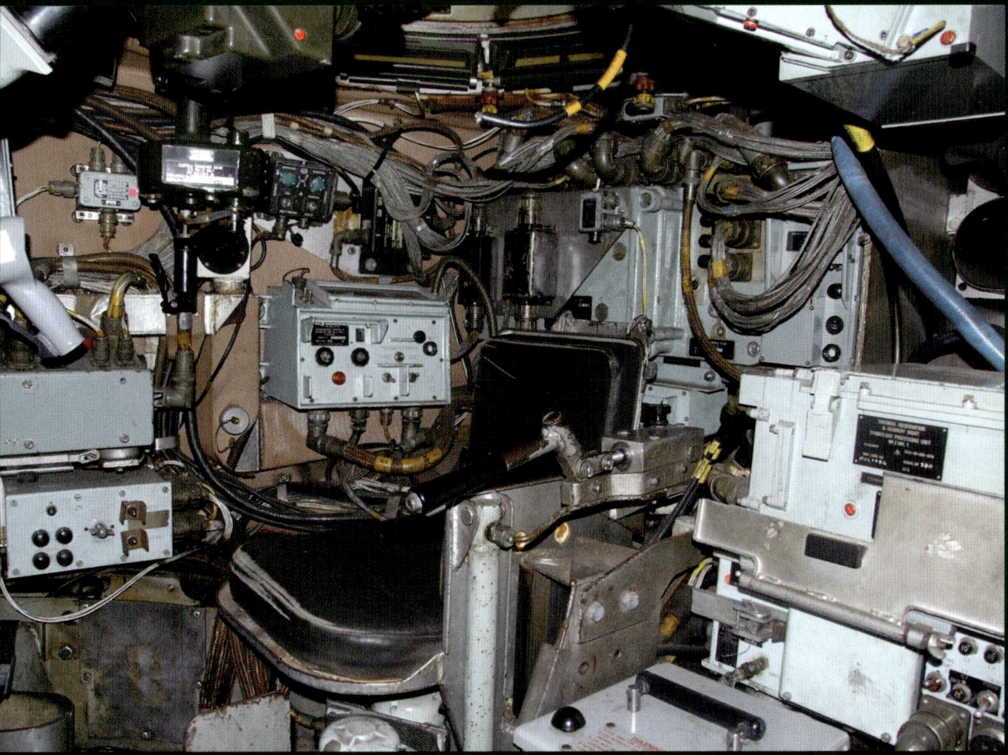

Showing how cluttered and very unergonomical the commanders area is compared to modern tanks such as the Challenger 2. On the left of his seat the black handle is used to raise his seat hydraulically, the box to his right controls the input of range corrections and will also indicate faults in the fire control system [RG].

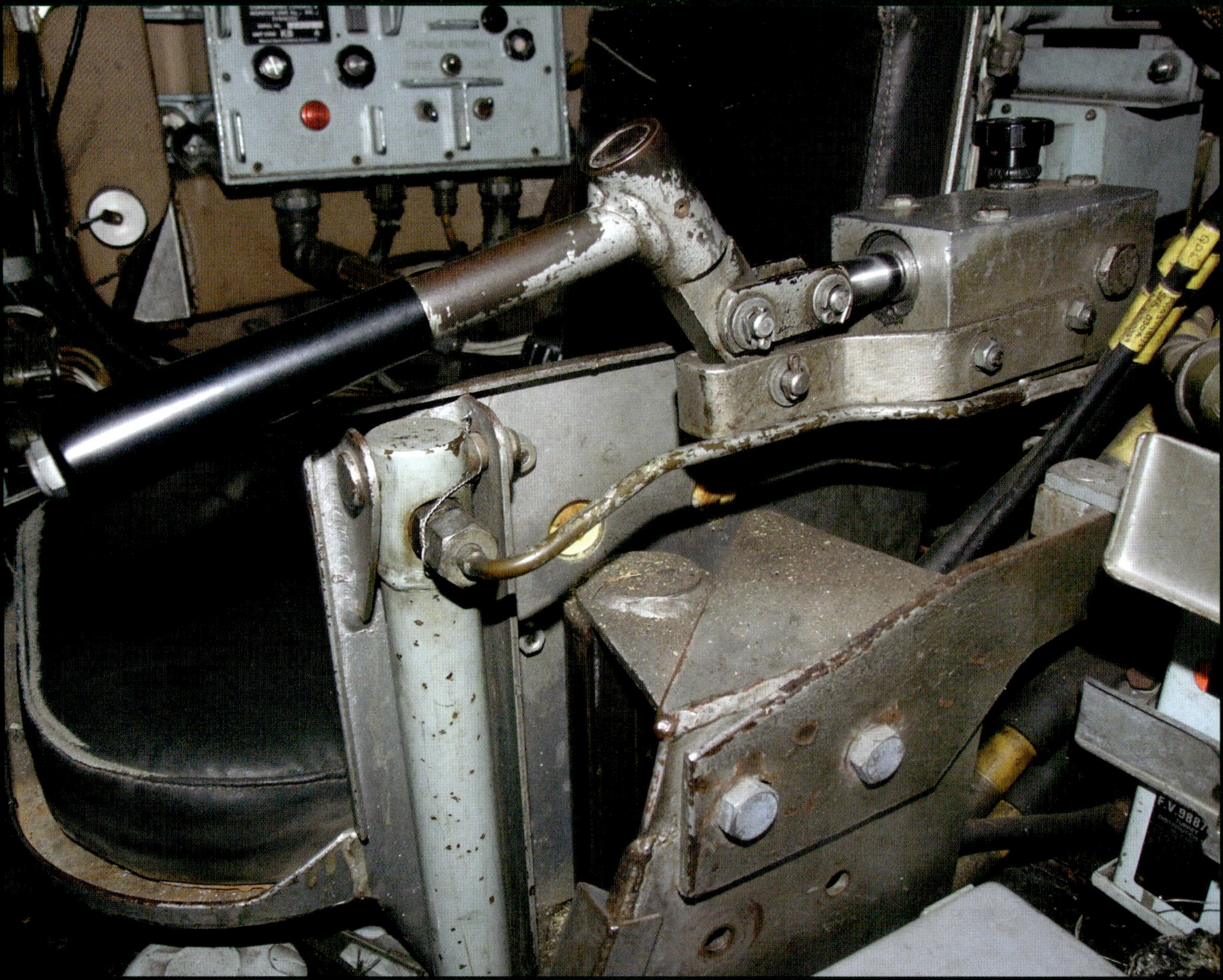

Close up of the commanders hydraulic pump handle for his seat, this enables him to raise the seat so he can sit "head out" in comfort, to lower it he simply unscrews the black knob visible at the rear [RG].

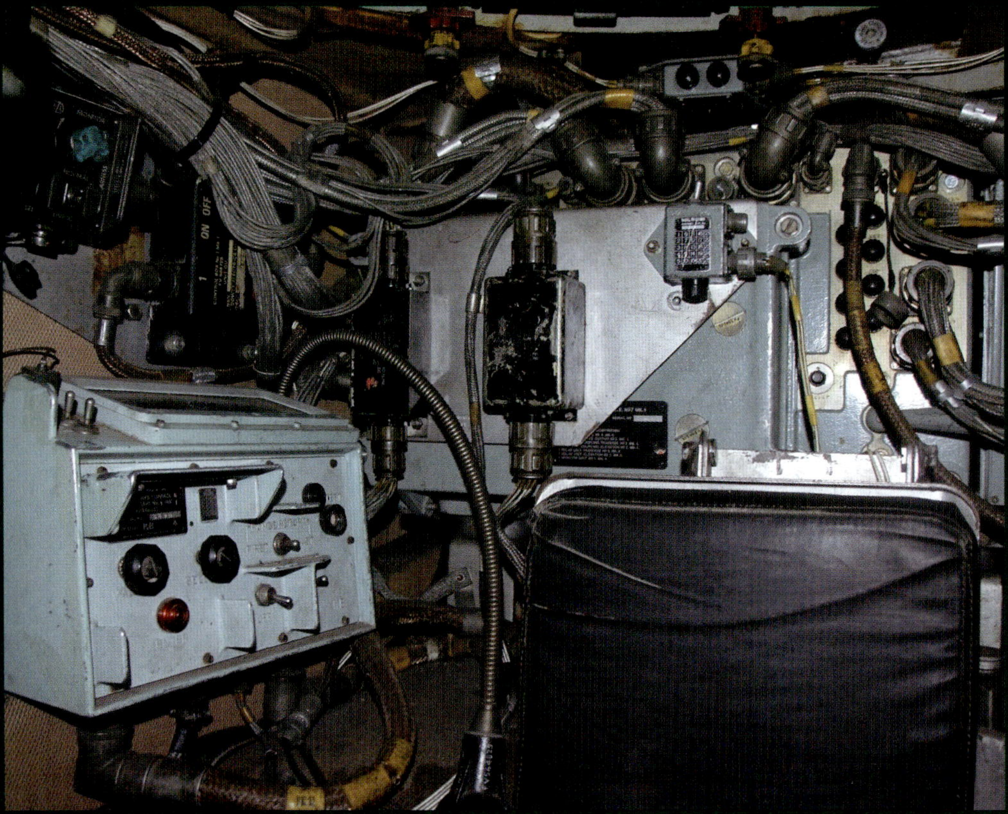

A mass of cables dominate this station to the left is the CCMU commanders control and monitoring unit, which is used for servicing and input for range data. The whole area to the rear is filled with the electronics used to provide power traverse and elevation [RG].

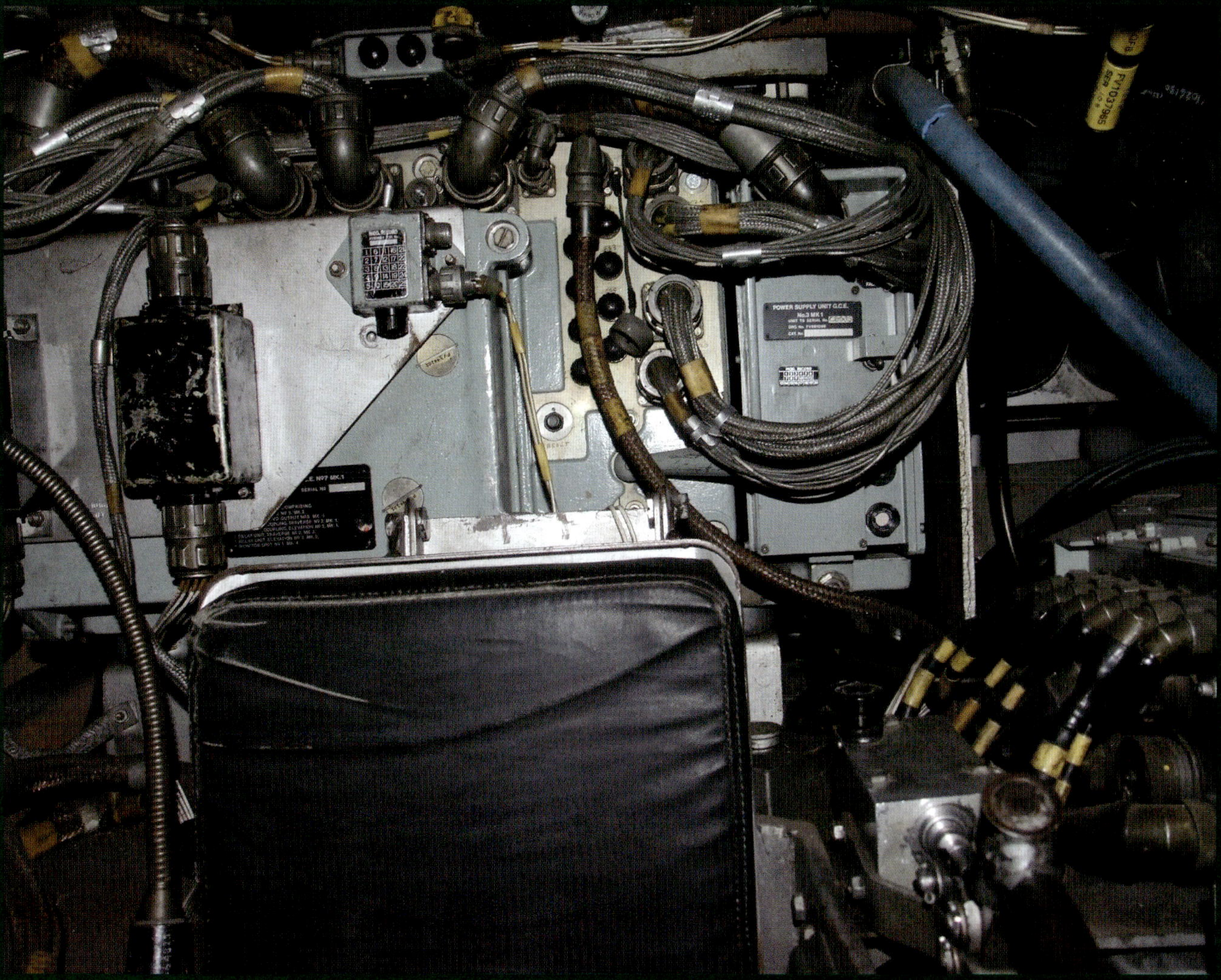

A shot showing the commanders seat back rest which can be flipped to a horizontal position and in conjunction with the hydraulic pump gives him a fairly comfortable seat to use while working opened up. The silver object to the left of the picture is the commanders map light on its flexible stalk [RG].

The Commander's control and monitoring unit, information is displayed in the window on the top and adjustments are applied using the various controls below, on the extreme right of the window can just be seen the commanders GPI gun position indicator, this shows him the position the gun is pointing and is used when fully closed down [RG].

The commanders control visible in this shot from left to right are the commanders gun control handle with its very small dome on the top, this dome is used to control the traverse and elevation under power. The box to the right and above is the Laser range read display and the large green control allows the commander to hand traverse the cupola [RG].

Some of the drivers many instruments, in this his rev counter and lighting switches, visible below is one of the hull batteries [RG].

A general view of the commander's position [RG].

View of the driver's position on a well-worn Challenger, visible in the centre of the picture is the right steering lever and behind that is the gear range selector, with one of the vehicle batteries seen behind [RG].

This shows the luxuries the driver had in the Challenger 1!! This is the outlet for the forced air system which could be linked to the heaters when fitted [RG].

Visible in the front of this shot is the right steering lever, to its right is the gear selector knob, behind it is another battery [RG].

A view of the gunner's seat. Mounted directly in front of the gunner is the Tank Laser Sight, the right eyepiece gives the gunner the x10 view while the left one shows the laser range. To the right can be seen the TOGS screen, and notice how ergonomically poorly sited it is. Below it is the elevating handwheel, with its firing switch in red visible on the left, central is the gunners fixed firing grip (which has a firing switch, armament selector switch and the laser/autolay switch), directly behind in the grey box are the controls for the BRS system, and finally on the right is the gunners power traverse controller (virtually the same as the commander's), with the hand traverse handle being the brown one behind it [RG].

General view of the gunners station, the large circular black wheel is the hand elevation, to the right is the gunners power controls similar to the commanders and to the right of that is the gunners TOGS screen [RG].

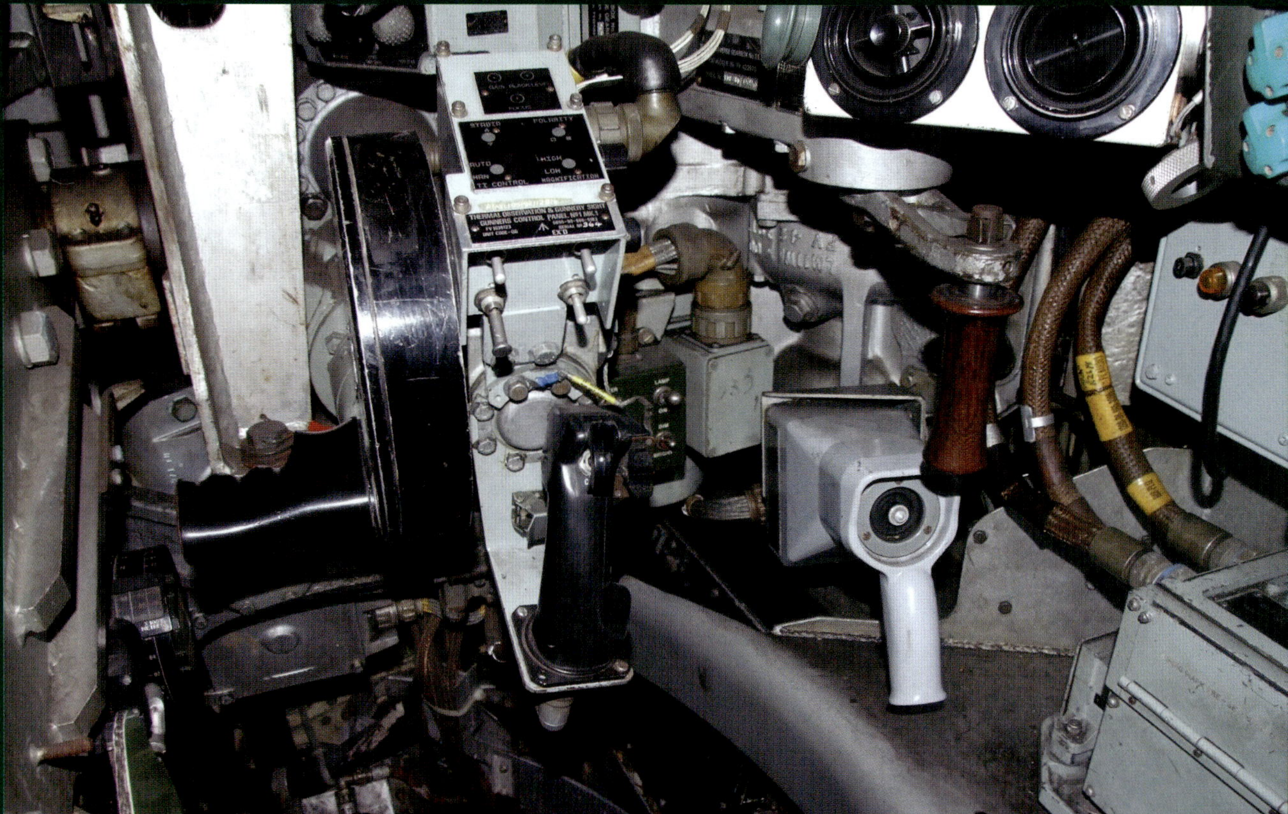

A good shot showing all the gunners controls, the large black wheel is the elevating handwheel and to its right are the controls for the MRS system; also there is the fixed firing grip again similar to the commanders. On the left is the power controls, and above them the brown handle is the hand traverse handle when not using power [RG].

The gunner can fire both the co-ax and the TLS by means of foot pedals, both of which are visible here with co-ax on the left [RG].

Close up of the gunners TOGS viewer, this can be swung out to allow slightly more comfortable viewing, above it is the control panel for the power laying and travers equipment [RG].

The gunners TLS with eyepiece for the aux sight vible to the right of the picture. The gunner views through the right eyepiece on the TLS and the lased range appears in the right eyepiece [RG].

DSCF0164/ The eyepiece of the auxillary sight is visible in this shot as are the controls on the TLS, the red button is another way for the gunner to fire the [TLS].

In this shot the silver box with the diamond tread is a vital part of the turret equipment, for it contains the crews' mugs and tea or coffee making supplies. The rods projecting out with round end caps are used to secure rounds in the rack above the "tea box" [RG].

A view inside the turret from the commanders seat. Visible on the turret left wall are the brackets for the operational ready APFSDS rounds; to the right is the co-axial machine gun, the 7.62mm GPMG. Dominating the picture is the main armament breech, the red bracket in the center is known as the Stop, Running Back, fitted during training. It is fitted, as its name suggests, in the event of all pressure to the recoil system being suddenly lost, when it prevents the gun from suddenly sliding to the rear. Such an eventuality could cause damage to the equipment, but more importantly, injury to the crew [RG].

Although this in-service vehicle also has no ammunition stowed, we can see the human touches more common to life on a tank. The two crewman's helmets are interesting in that the AFV helmet on the left has been touched up with black paint, probably more in humour than as a real effort at camouflage [S. Park 2RTR].

Looking down the open loader's hatch on a 2RTR Challenger 1 [S. Park 2RTR].

Looking at the rear of the turret; from left to right we have the commanders seat, and visible on the left of it is the handle used to raise and lower it. In the centre is the Boiling Vessel and above that is the main turret electrical distribution box with the Symbology Processor for the TOGS system located on top of it. To the right can be seen on VRC353 VHF radio fitted below the projectile racks. Various radio fits could be installed, depending upon the role of the tank [RG].

The GPMG co-axial machine gun's location in which can be seen the feed chute for the disintegrating link ammo belts, and below the chute for spent case and links. The spent cases would be fed into a larger container located on the turret floor. In the centre of the picture is the pneumatic system pressure gauge, which must be at 500lb pressure before it is safe to fire [RG].

This is the stowage for the 7.62mm ammunition, which would be removed from its containers and linked into several long belts and stowed in these trays. Lying on top are two vent tube magazines with a third stowed correctly [RG].

A left front view of the loaders station showing the APFSDS stowage on the forward face of the turret; and behind those, on the turret roof is the loaders drop down periscope. It is raised flush with the roof when he is loading the weapons [RG].

The 120mm gun cradle with the GPMG visible on the left; on top can be seen part of the recoil system. The large red painted component is the Stop Running Back, secured at the end of the gun cradle and to the breech ring. It prevents the gun running back in case of a loss of pressure [RG].

Loaders position showing in the centre the vertical tube holding the air vent outlet, to the left is projectile stowage, with above it the large green control box is for the Clansman radio system and above that is the loaders safety switch. When switched to safe it will stop all power traverse and elevation dead, unlike the commanders [RG].

Visible here are the feed tray for the GPMG (top) and the link exit chute (below). The green solenoid is for the electrical firing of the GPMG, and the pressure gauges tell the crew if less than 500psi remain in the recol system. The orange topped handle is the breech closing handle, and just behind that is the breech opening handle [RG].

Close up of the loaders drop-down periscope, this can be raised flush to the turret roof when he is loading; it is retained by means of the knurled knobs visible on either side [RG].

Located below the APFSDS rack is the large ammunition container for the GPMG. Several belts of 200 rounds apiece are linked together and stored in this bin. To the right rear can be seen the vent tube magazine in it's stowage location, normally six would be carried each containing the .625 vent tubes which when fired electrically ignite the bag charge in the 120mm gun breech [RG].

Central in this photo is the ammunition stowage with a vent tube magazine posed on top , to the left the small stowage bin is known as the "brew bin" in which brew kit and tea mugs are kept. Behind the vent tube magazine is the white painted control panel for the NBC pack [RG].

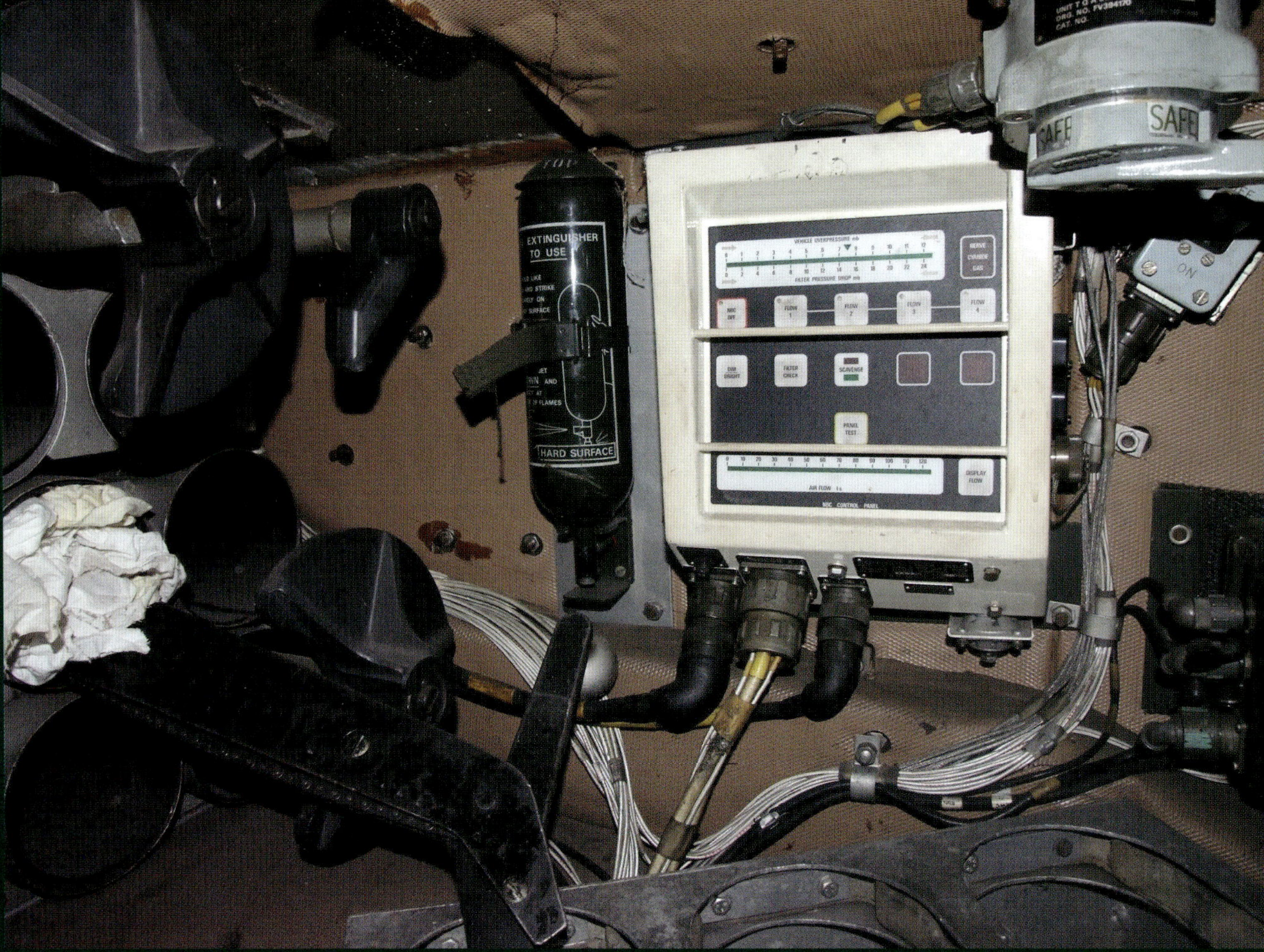

A good view of the NBC pack control panel, the object stuck into the horizontal projectile stowage is the breech opening tool, used when carrying out maintainance on the gun [RG].

Horizontal projectile stowage located above the radio tray, although this vehicle has one Clansman VRC 353 VHF radio fitted a normal gun tank in a troop would carry two of these, a command or control tank in RHQ or SHQ would have these plus a VRC 321 HF set, which would reduce some ammunition stowage [RG].

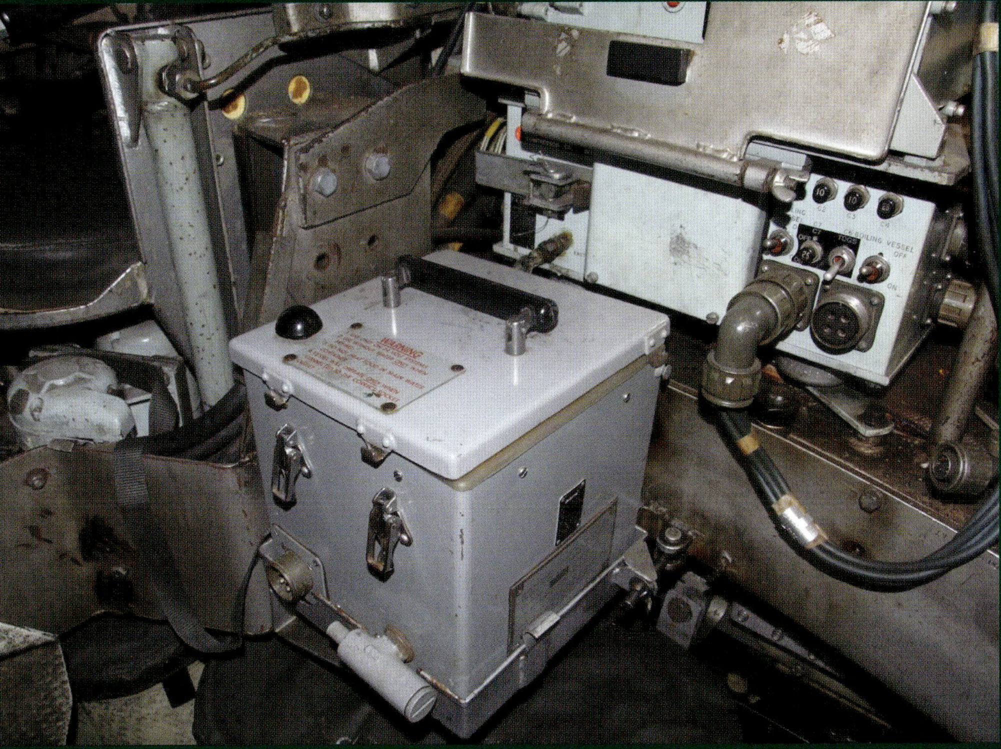

No book on a modern British AFV would be complete without that most important of items to the crew, their boiling vessel. This gives the crew a contant supply of hot water and allows them to cook their rations which these days are of the "boil in the bag varity", making life a lot easier [RG].

To the left of the picture can be seen the 3 round ready bin for bag charges. This could hold three APFSDS charges or six HESH/SMOKE charges. Centrally can be seen the breech closing and opening levers and the handle at the extreme right is used by the loader to pull a shield down, making the circuit to fire the gun preventing the loader from falling behind the gun as the vehicle moves. Many Challengers were fitted with the "Chase" modification named after Staff Sergeant Charlie Chase of the 4/7th RDG. This modification closed the breech when the safety guard was pullled to the rear, thus saving vital seconds in the loading process [RG].

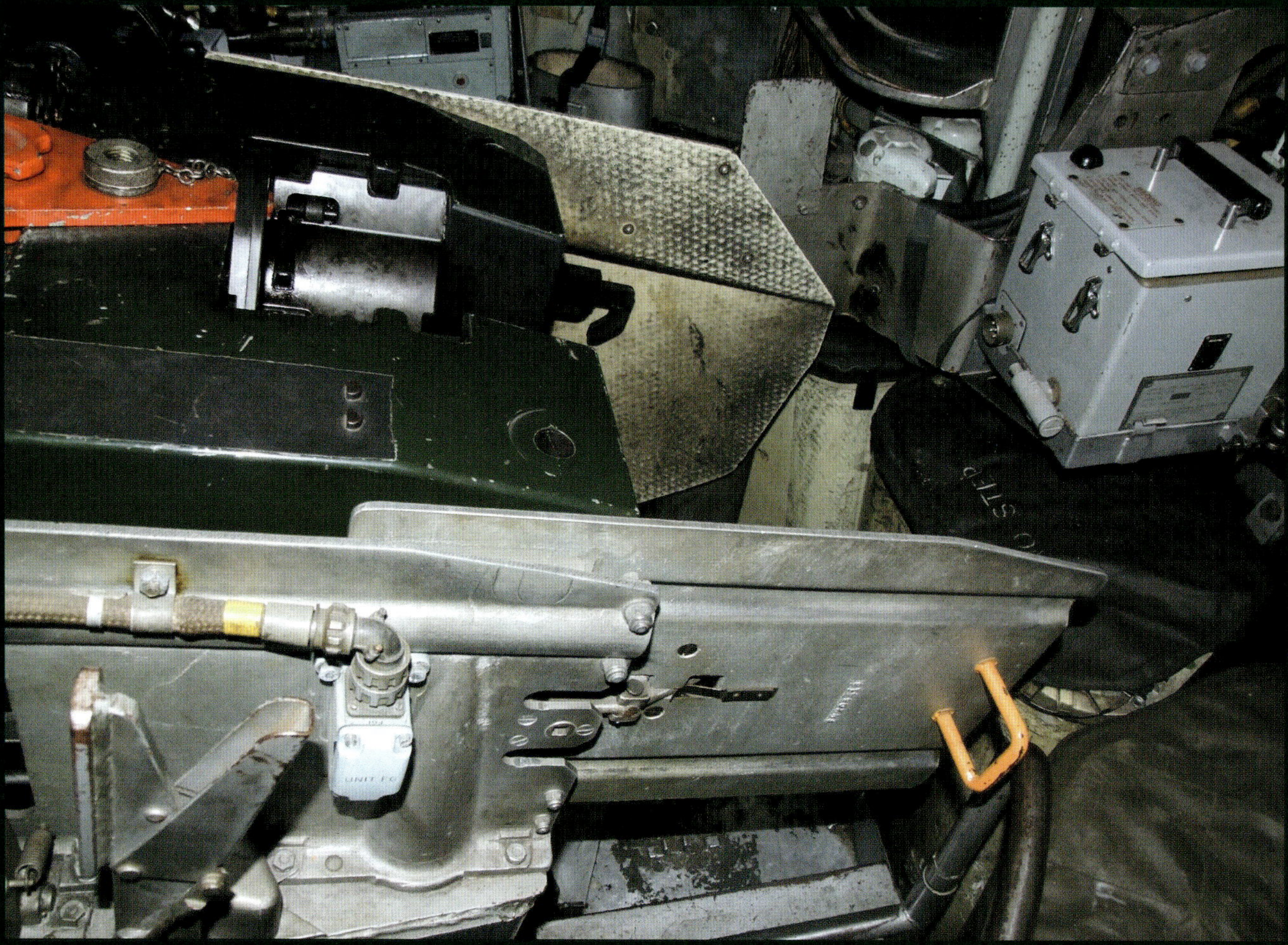

This shot shows the loaders guard pullled to the rear, when the gun recoils it breaks a cam and the guard is released thus giving the loader room to move to load the gun [RG].

The large circular object in the centre of the turret floor is the Rotary Base Junction, this allows the the transmmision of power from a fixed object (the hull) to a movable one (the turret), its a very advanced form of slip rings but works remarkably well [RG].

One of the many junction panels located throughout the tank. Notice the use of very heavy plastic shielded or copper armoured cables to protect them from wear and tear [RG].

This shows two of the three charge bins located at the rear of the fighting compartment, when not in use they have a plastic cover fitted to minimise the ingress of water and dirt [RG].

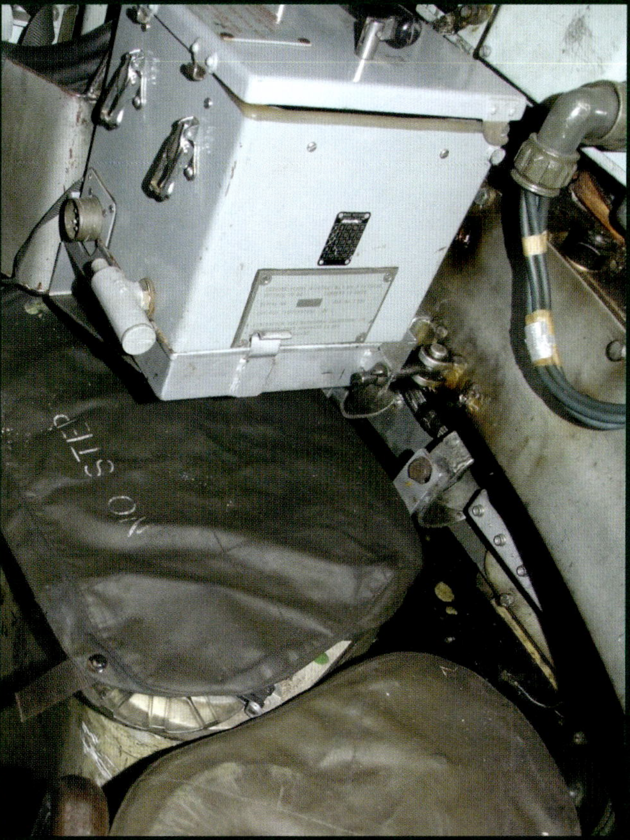

All three bins can be seen in this shots and the loaders safety rail is in the foreground, when fitted his removable set is located onto this rail [RG].

A good view of the Breech Opening Tool (BOT): the large end locates on the bottom right of the breech ring and the top goes to the left, then to open the breech the handle is wound until the breech drops to the loading position. Until the introduction of the Breech Opening Handle this was the only way to initially open the breech for the first time, subsequently the breech opened automatically during runout after firing [RG].

A rather pristine looking interior to the brew bin, after a few days in the field it will collect all sort of odds and ends, all of which the crew will swear are vital [RG].

Two parts of the recoil system can be seen here, the large cylinder is one of two recoil buffers, the other being located below the gun, the other cylinder is known as the replenisher and it does as its' name implies: it keeps the buffers topped up with oil [RG].

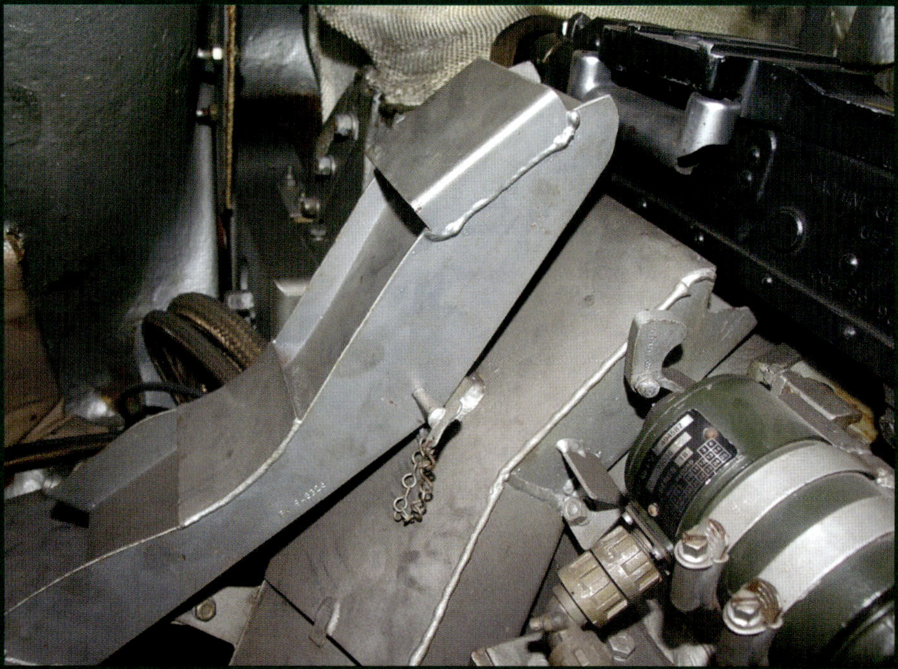

A close up of the feed tray to the GPMG, this allows feed of the ammuntion to be relativley smooth with no obstructions, directly underneath is the link exit chute, which takes away the discarded links that have held the rounds in the belt [RG].

These strange looking objects are the projectille retaining clamps, designed to be an improvement on the type found on Chieftain [RG].

A crewmans nightmare, the internal dust cover. Designed to stop water, moisture and dirt entering the turret, they only seemed to fit once and that was when the vehicle was built, replacment ones came with no holes provided and as can be seen it is not the easiest location to work [RG].

Visible here is the loader's safety switch, the loader's turret light could be dimmed by means of the lever on the right and the box to the right of the light is the gun ready light, which glowed red when all circuits and safety checks were made allowing the gun to be fired [RG].

This shows how the bazooka plate were located and secured, also visible is the most useless fitting on a MBT-the mirror arm, hardly ever used and normally found in this position [RG].

This is one of the fire extingusher handles located on the tank, simply twist and pull and the engine compartment will be filled by the extinguishers... hopefully putting out the fire [RG].

The forward stowage clamo for the tow rope, notice how the head of the rope is protected [RG].

Several of these circular access panels are located along the hull top, removing them allows access to the flexible fuel panniers [RG].

/ It can be seen in this shot how the fire extingusher handle is protected by its recessed location. The white panel tells the user how to operate the extinguisher in an emergency, although I am not too sure if the user would have time to read that pannel in such a case! [RG].

The tow rope stowage bracket half way along the hull on the left side [RG]

Looking along the left side of the hull showing the bazooka plate fixings and tow rope brackets, as this vehicle did not go to the Gulf it does not have many of the extra brackets that were fitted for the up armouring [RG].

The small brass tube on the tow rope has details of the rope stamped into it, this aids in servicing the rope as the user can see how old the rope is [RG].

The rear end of the tow rope stowage before the rear cover, the clamp and wing nut bracket secure the rope when not in use [RG].

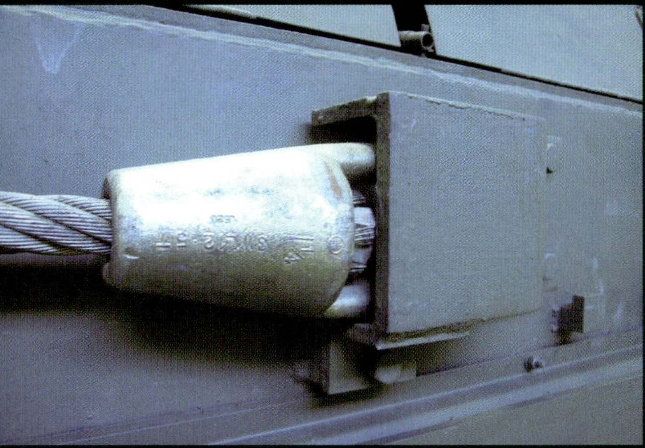

The rear mounting for the tow rope again notice how well the head of the rope is protected by the bracket [RG].

This shows the method of securing the tow rope bracket to the hull side with two bolts. Thus making it easy to replace if damaged [RG].

Aboard railway flat wagons, a good rear view of a Challenger 1 (most likely an unmodified Mk.3) of The Royal Hussars in the early 1990s. Notice the high visibility markings legislated for road use in Germany and the large box fitted on the left of the gun crutch [RG].

Left side view of the gun clamp mounting bracket, visible is the release catch that will allow the clamp to be swung into the vertical position [RG].

A view of the right hand rear lighting, on the far side is a reflector, in the centre is the normal army red light lens and the green object next to it is the light for the number plate [RG].

A view from the side showing how the relector is mounted, and below the lights the very small rear mud guard [RG].

Right rear view which shows the armour joins very well and also the small rail that is located on both rear hull corners [RG].

Just visible beyond the rail is one of the very flat fuel filler caps located on the rear hull [RG].

Right rear tow rope stowage point and directly above are the empty brackets for the pioneer tools normally carried [RG].

The right hand exhaust, again looking very clean, if required a thermal shorud can be fitted over both exhausts to minimise the IR signature [RG].

The corner of the rubber resilliant rail protecting the gun barrel, and poking out between the rails is the fuel tank breather pipe [RG].

A close up showing how the tow rope head fits into the front stowage location [RG].

Notice how the tow rope is secured in this shot compared to the left side.poking out between the rails is the fuel tank breather pipe [RG].

The right hand fire extingusher handle, there are two extinguishers outside and two inside the driver's cab [RG].

The idler wheel was adjusted by means of a large nut on the front of the idler mounting bracket to regulate track tension [RG].

Each sprocket was secured by 13 large bolts [RG].

An instructional aid showing how each road wheel station was attached on the Hydrogas suspension system employed on the Challenger [RG].

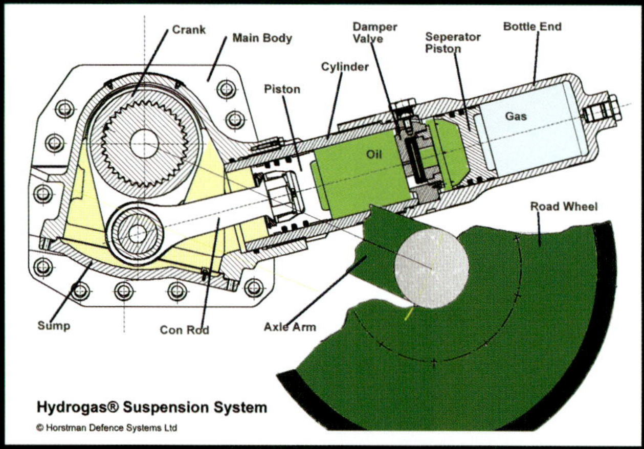

A sectioned view of the hydrogas units giving a good illustration of how the system works, another advantage is that as each unit is mounted separately. In the event of mine damage it was relatively easy to replace [Horstmann].

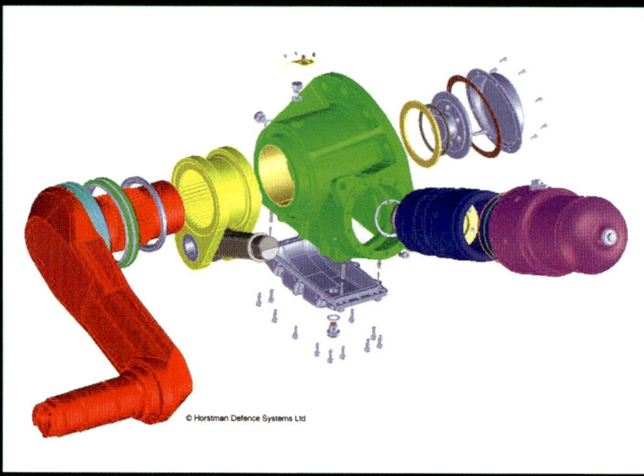

A colourful exploded diagram of the hydrogas suspension units as used on the Challenger, when shown like this it can be seen that the system is relatively simple [Horstmann].

The hydrogas suspension units as normally seen in standard green paint, use of this system helps keep the vehicle overall height down compared to a torsion bar equipped vehicle, which will always be higher due to the torsion bars running across the hull bottom [Horstmann].

This show how the hydrogas units are fitted to the hull sides on the Challenger. What is interesting in this shot is that this particular vehicle is one of the early prototypes, so the mounting holes for the original Chieftain type Horstmann suspension system once fitted are still clearly visible, not present on production hulls [RG].

This cover is located on the top of the TOGS barbette and allows access to componets for maintenance [RG].

A close up picture of a Challenger 1 on the firing ranges. Several interesting details can be seen; the five tubes with rubber covers on the turret front plate are the multi barrelled smoke grenade dischargers. These are fired by the commander to give instant local smoke cover for the vehicle and are especially useful against antitank guided missile threats. The Thermal Image Sight Head can be clearly seen in the TOGS barbette, and above it is the commander cupola, which is flying a green flag (which signifies the tanks guns are unloaded during a range period). The large mast in the background is only fitted during range periods, and it transmits live images to the range control tower of all aspects of the crew drills, allowing an accurate debriefing after the shoot [RG].

The front door of the TOGS barbette, which protected the Thermal Imagery Sight Head when the TOGS system was not in use [RG].

The access pannel recessed into the top plate of the TOGS barbette, seen from the turret roof [RG].

Cooling louvers for the TOGS system [RG].

The TOGS Barbette also had an access pannel secured by two large Allen bolts on its rear side face [RG].

The bracket supporting the lower edge of the TOGS unit [RG].

The right hand smoke discharger, seen from the left. Each discharger unit held launcher 5 barrels and would be used to generate an instant smoke screen [RG].

Viewed from the turret roof, the smoke discharger used a fixed rubber cap on each barrel to keep rain out. The caps would be blown off when the dischargers were fired [RG].

The front right of the turret showing the small tie down brackets that are located all over the turret and just at the bottom of the picture is the mounting for the smoke discharger [RG].

The pannier on the left front of the turret was fitted with a rubberised plastic cover, and the layout of the pannier was carefully configured not to interfere with the smoke discharger mounted on the left turret front plate [RG].

The front right of the turret, showing how the main gunner's sight is recessed into the turret roof, the smoke discharger and the gun mantlet with its bellows. Below the gun we can see the open driver's hatch [RG].

The turret seen from the front left. The loader's periscope, his split entry hatch and the commander's cupola are all visible on the roof of the turret. The depth of the left side stowage basket (a retrospective but standardized modification) can be seen [RG].

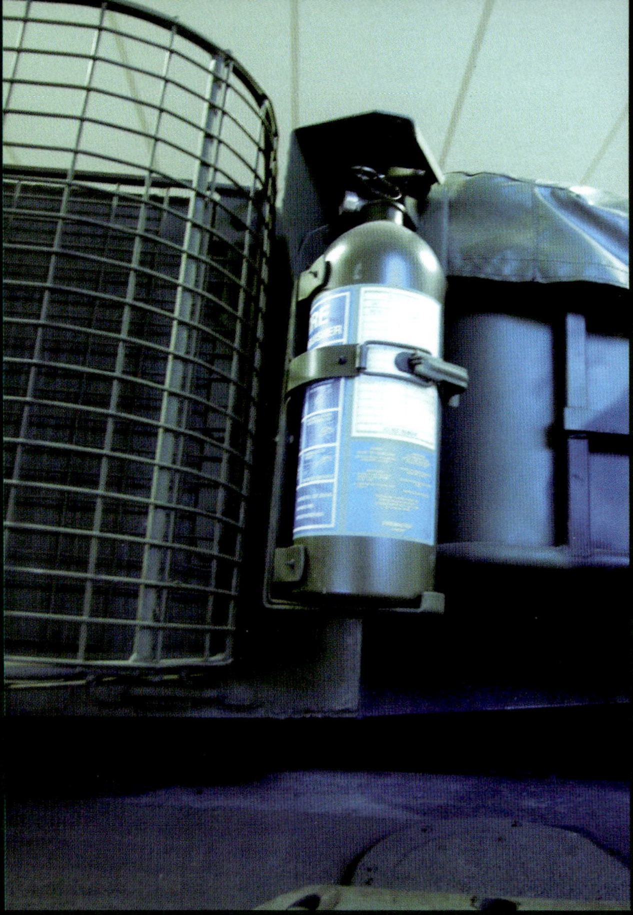

The lower edge of the left turret side stowage basket follows the turret lower edge as can be seen here [RG].

On the left side of the turret towards the rear, one of the vehicle fire extinguishers was stowed on a bracket. Its blue label can be seen distinctly [RG].

Even the factory layout for the Challenger 1 Mk.2 incorporated stowage racks for water jerricans, which would quickly multiply after delivery to the armoured regiments. Seen here is the jerrican rack on the turret rear stowage bin [RG].

A closeup three quarter rear left view of the Challenger 1 Mk.2 at Shrivenham, we can see the exhaust port, fuel tank breather tube, towing hawser stowage and grease gun stowage on the turret rear [RG].

Here is the fire extinguisher bracket seen without the extinguisher mounted from the right side of the turret [RG].

The turret casting number, similar ones are to be found on most AFVs, from them a lot of detail on the vehicle's history can be obtained [RG].

Right hand side of the turret with part of the cupola visible, the mesh stowage basket, and further forward the TOGS barbette [RG].

The plastic cover over the right hand turret basket, with just forward of it one of the Clansman aerial bases, the half oval bracket on the turret rear is a guide for the lifting sling when the turret is being lifted off or on to the hull [RG].

The rear of the commanders cupola, and behind it the pressure relief valve [RG].

The very large and heavy cover in this shot covers access to the gun power equipment on the rear of the turret roof [RG].

Under the large hatch can be found the two turret batteries and a funnel for the internal turret water tank, the smal circular raised object next to it is a breather for the batteries [RG].

A top view of the turret battery access hatch, showing the lifting handles and between them the two locking keyways [RG].

Left rear Clansman antenna base, and next to it a fixed mounting for the rotary amber beacons used when on public roads [RG].

To the left of this shot can be seen the loaders periscope, it is actually quite suprising how much the loader can pick up using this sight. Central is one of the two hatches for the loader, notice the leather strap to help close it and the green protective padding [RG].

The rear of the two loaders hatches showing quite clearly the two locking handlse used to secure the hatches when closed down [RG].

The plastic cover over the front basket, if covers are not supplied crews will produce the most ingenious versions themselves [RG].

Overhead view of the right hand side turret mesh basket, although not substantial every little bit of stowage helps [RG].

The forward edge of the turret side stowage basket [RG].

The left hand rear stowage basket which likes it opposite number on the right has a factory produced plastic cover, secured by straps [RG].

Looking rearwards can be seen the large rear turret stowage bin, with the grease gun to its right and a general view of the transmission decks[RG].

Close up of the turret bin showing the large hinges, this is a part that is incorrect on the Tamiya 1/35th scale model [RG].

Central shot of the rear bin again showing the hinges and the single water jerry can located on its rear face [RG].

A view of the rear turret stowagebin with the turret lifting strap guide between it and the right side turret stowage bin [RG].

The grease gun stowage on the left rear of the turret. The Challenger is a big tank and requires a huge grease gun! [RG].

A heavily camouflaged Challenger 1 MBT crosses a pontoon bridge built by the Royal Netherlands Army Engineers during one of the Cold War's large exercises in West Germany. Unlike its Soviet and German contemporaries, the Challenger 1 design did not incorporate deep fording capability [DB].

The top view of the grease gun stowage. The Grease Gun would be used for crew mainance both in camp and in the field. The top is removed and a full tub of grease inserted... and away you go [RG].

Looking forward along the left side of the turret, showing the very pronounced slope of the turret top [RG].

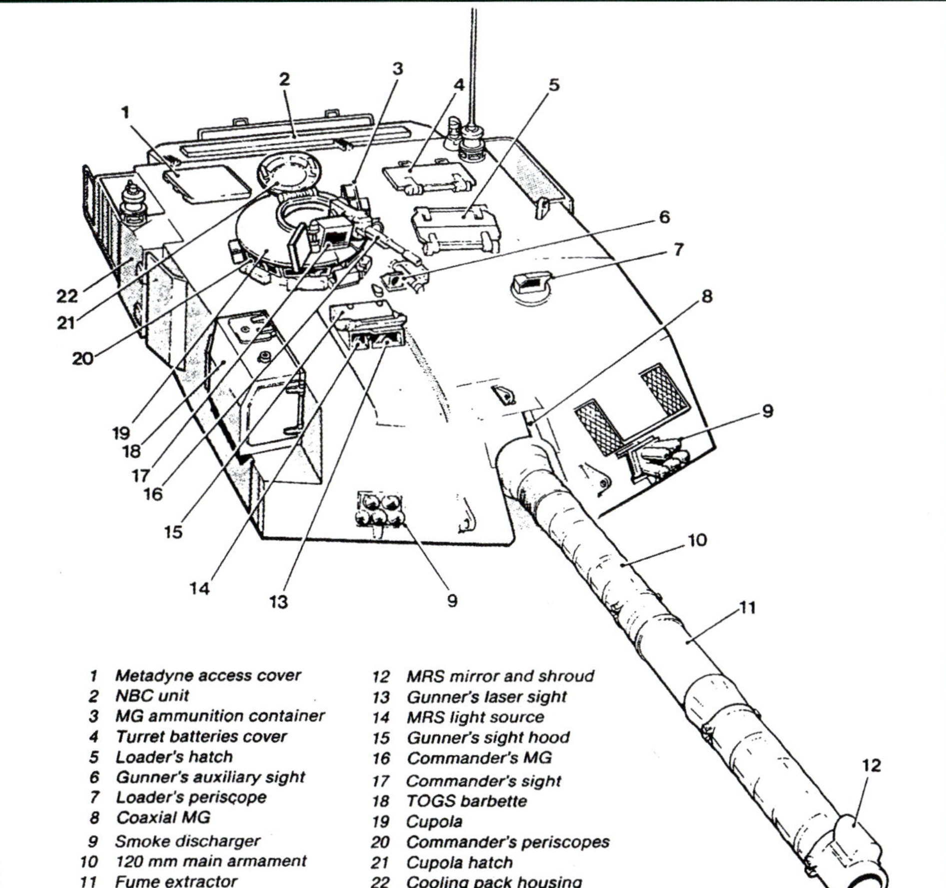

A manual diagram showing the basic layout of the Challenger 1's turret [Crown Copyright].

A front three quarter view of the Shrivenham Challenger 1 Mk.2, in the Royal Military College of Science collection [RG].

The Challenger 1 Mk.2 at Shrivenham seen from the front. The mass of Chobham armour protecting the turret front is evident. The narrow mantlet and trunnions inherited from the original Chieftain's pointed turret front were cited as one of the reasons why the Challenger 1 was a less accurate gun platform than with the wider mantlets seen on the Leopard 2 and M1 series, and this feature was not inherited by the Challenger 2 design [RG].

The Challenger's hull glacis is a more complex arrangement than that seen on the M1 or Leopard 2 series MBTs, mainly because the Challenger inherited the Chieftain's central driver's position. Large Chobham armour packs were fitted in front of the driver's hatch and on each side of his position to give the top of the glacis plate excellent protection against hollow charge and kinetic energy projectiles. The basic layout was optimised for fighting hull down and the lower front plate was not Chobham protected. The hull design on the Mk.2 incorporated wet stowage for bagged charges on each side of the driver, low down in the hull. The driver had a good view forwards through the wide periscope, but even when driving head out, he naturally required input from the commander for directional changes and of course when reversing! [RG]

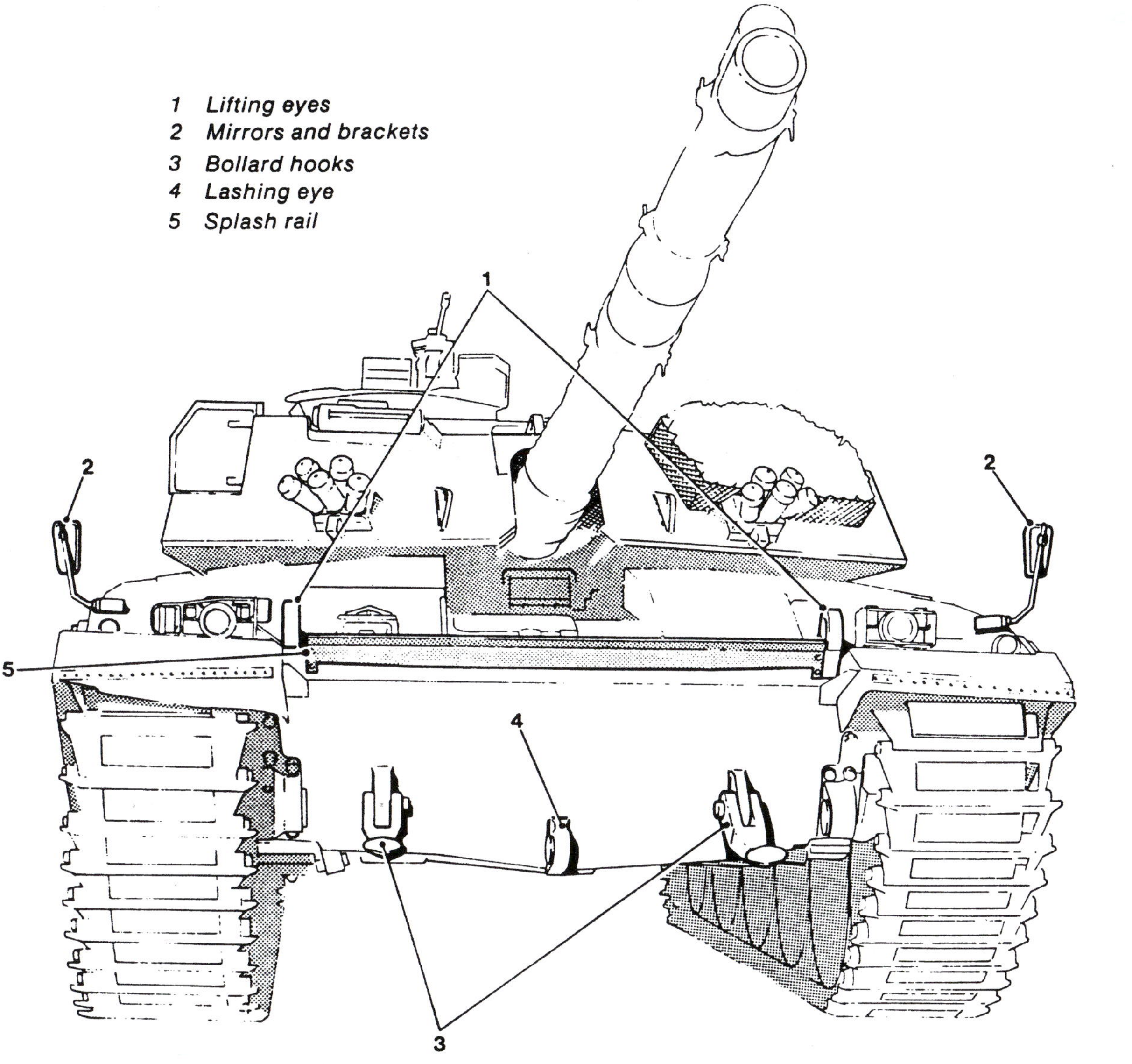

The Challenger 1 had to be secured for transportation by rail and by road, and this front view from the manual shows the attachment points that would be used to chain the vehicle down or to recover the Challenger's 62 tonne mass [Crown Copyright].

Changing the final drive on a Challenger 1. The track is visible, hanging from the rearmost return roller. The final drive is being hoisted away with a special lifting jig [W. Dunbar].

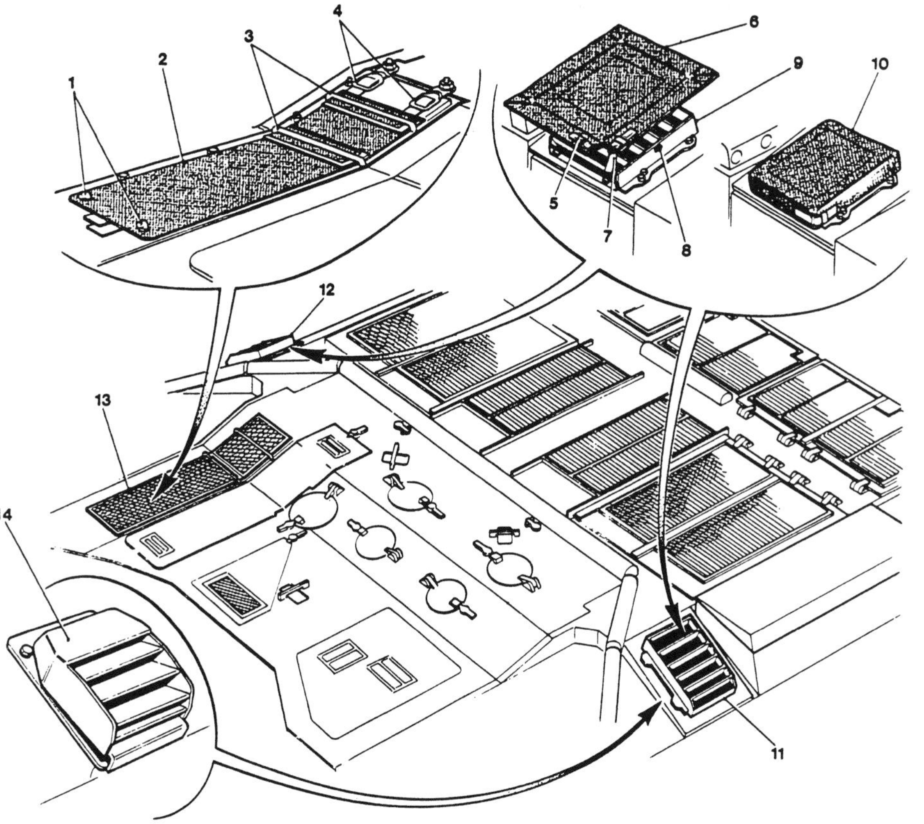

The air intakes, which were hidden under the turret bustle and the exhausts on each side of the Challenger's engine deck were provided with covers for use if the vehicle was being transported or stored. In this manual drawing we can also see the two types of exhaust outlet seen on the Challenger 1 Mk.2. [Crown Copyright].

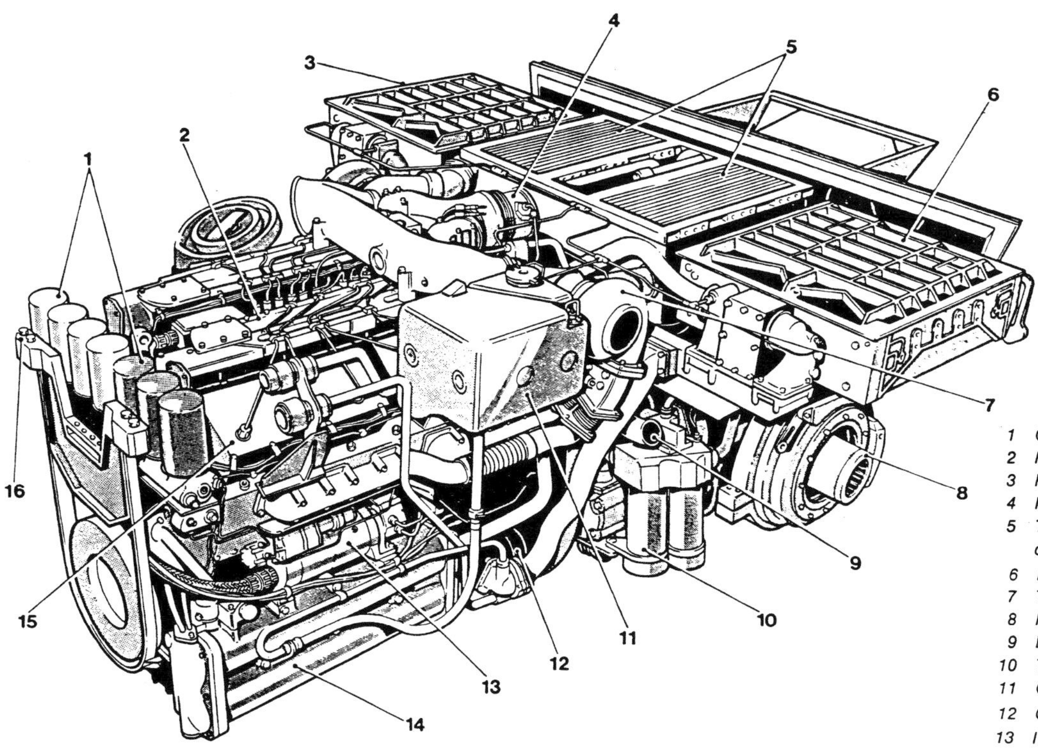

A manual drawing of the 1200 HP CV-12 engine which powered the Challenger 1 MBT. The size and weight of the Challenger power pack was a large impetus for the development of a dedicated Challenger based Armoured Recovery Vehicle, but for the first years of the Challenger's service recovery was performed by the Chieftain ARV [Crown Copyright].

The Challenger 1 Regiment would typically have been composed of 4 squadrons of MBTs and a Regimental Headquarters. One order of battle seen late in the Challenger 1's Career was the Type 50 Regiment, seen here with typical wireless call signs indicated

Type 50 Regiment RHQ	**Armored Squadron** Squadron Headquarters 3x Challenger 1 MBT 1 MBT	**Number 1 Troop** 3x Challenger 1 MBT	**Number 2 Troop** 3x Challenger 1 MBT	**Number 3 Troop** 3x Challenger 1 MBT	**Supporting Troops** (In Each Squadron)
	0B	**10**	**20**	**30**	**Admin troop** 1x FV432 Ambulance
4x Sultan CVRT 1x Spartan CVRT 1x 1-ton FFR	**0C**	**11**	**21**	**31**	**Light Aid Detachment REME** 1x FV432 24A 1x Warrior Repair 24B 1x CHARVV 24C 1x Warrior Recovery 24D
HQ Squadron Recce Troop 8x Scimitar CVRT **Light Aid Detachment** 1x CHARRV 4x Armoured Squadrons per Armoured Regiment (typically designated A,B,C,D)	**0D**	**12**	**23**	**32**	**Echelon Troop** 1x Samaritan CVRT 43A 2x Land Rover 2x 4-Ton 2x 8-Ton

The Challenger MBT program, for the first time in the Royal Armoured Corps' history, included a provision for a dedicated driver training tank, a practice long followed in the Bundeswehr and in the French Army amongst others in NATO. The vehicle was provided with a heated cabin which was laid out to carry an instructor as well as other student drivers, and the instructor could teach from mistakes driving across country or on base roads. He could also introduce faults to test driver reactions. Drivers could be changed at the instructor's discretion and the concept was successful enough to merit the purchase of Challenger 2 driver training tanks after the Challenger 1's replacement. The vehicle pictured sits derelict after a busy service life [RG].

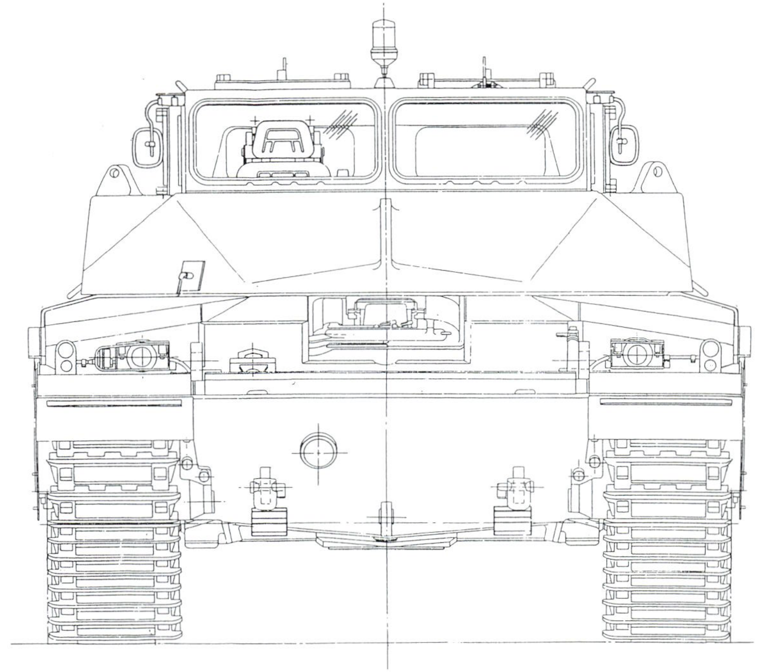

Challenger Training Tank – front and side view [CTT]

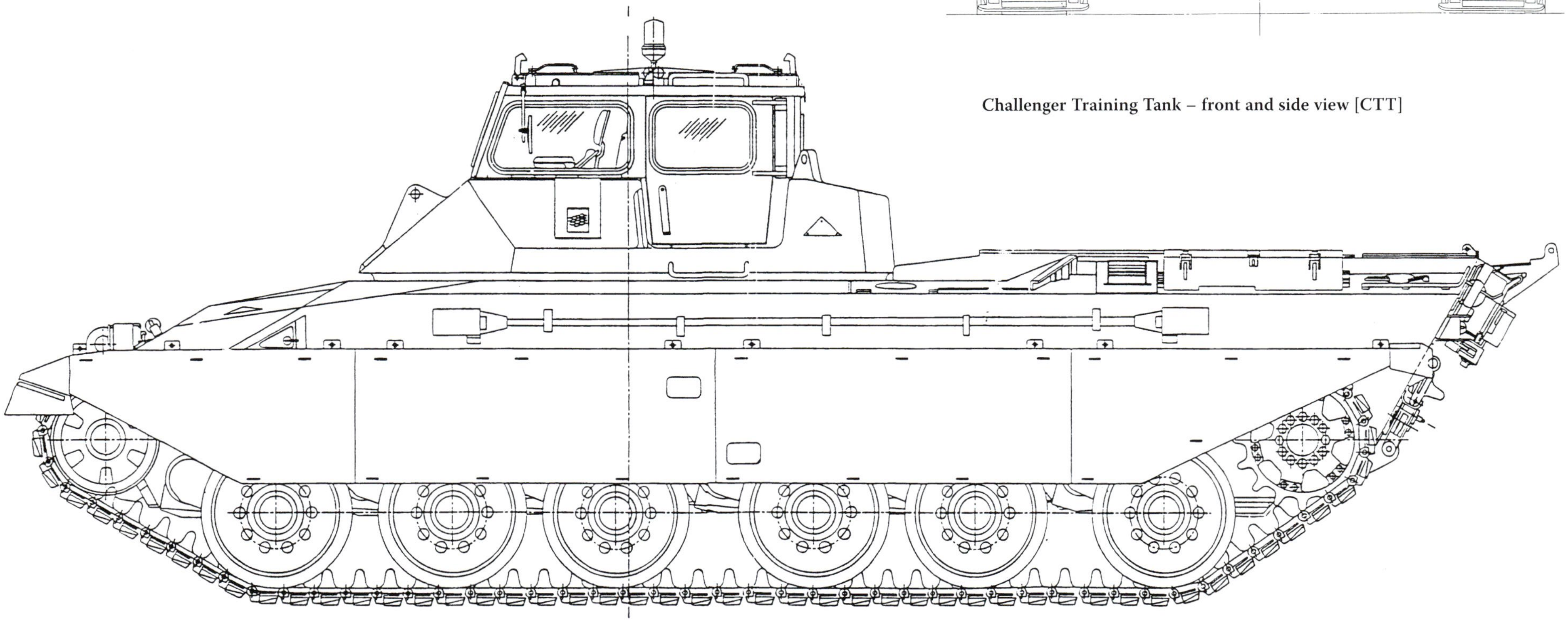

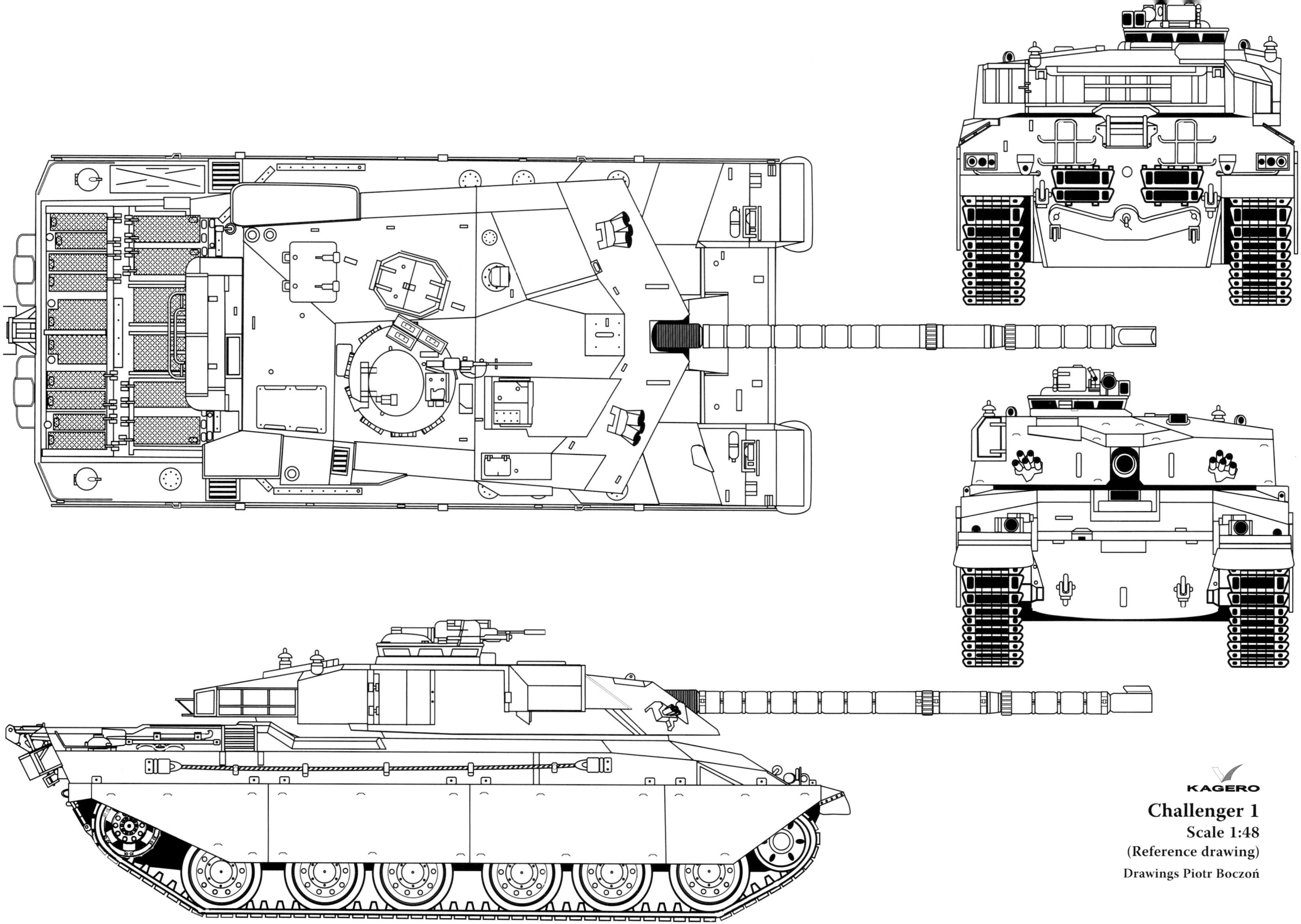

KAGERO

Challenger 1

Scale 1:48

(Reference drawing)

Drawings Piotr Boczoń

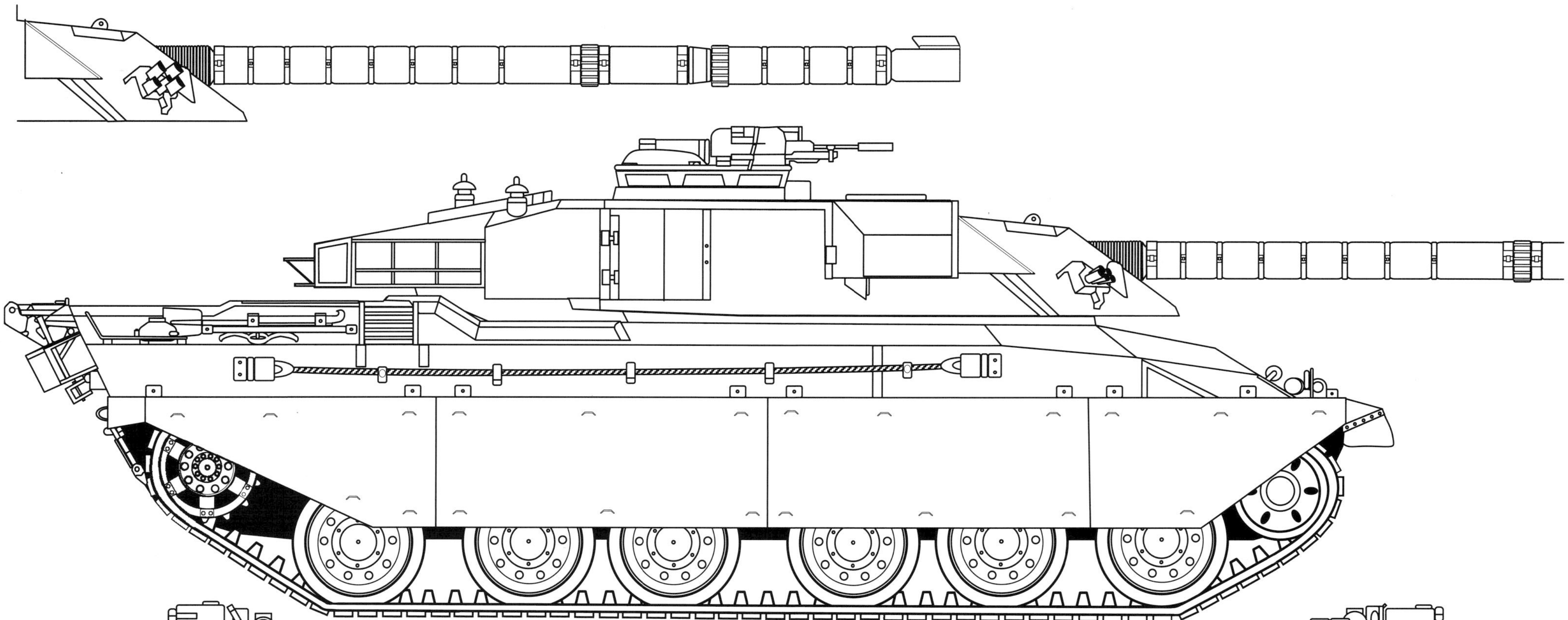

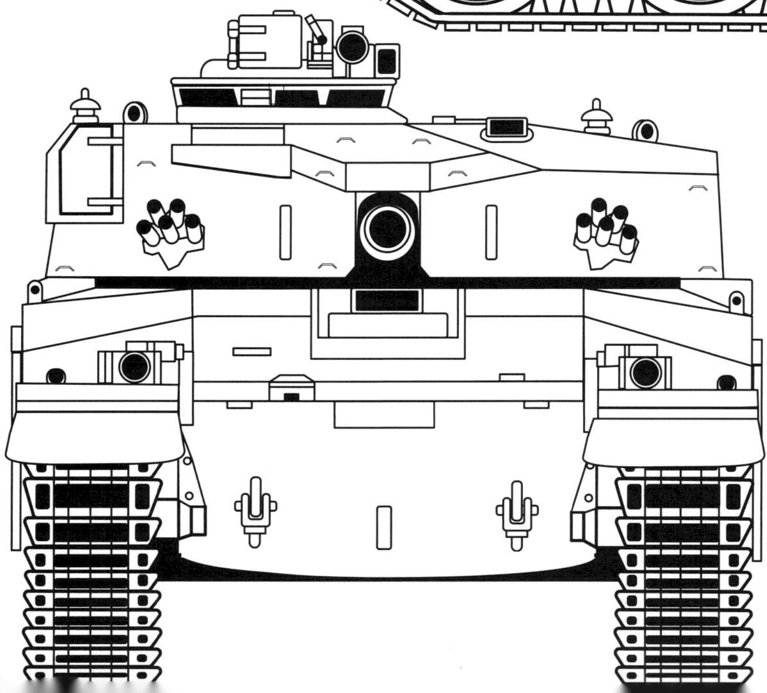

KAGERO

Challenger 1

Scale 1:35

(Reference drawing)

Drawings Piotr Boczoń

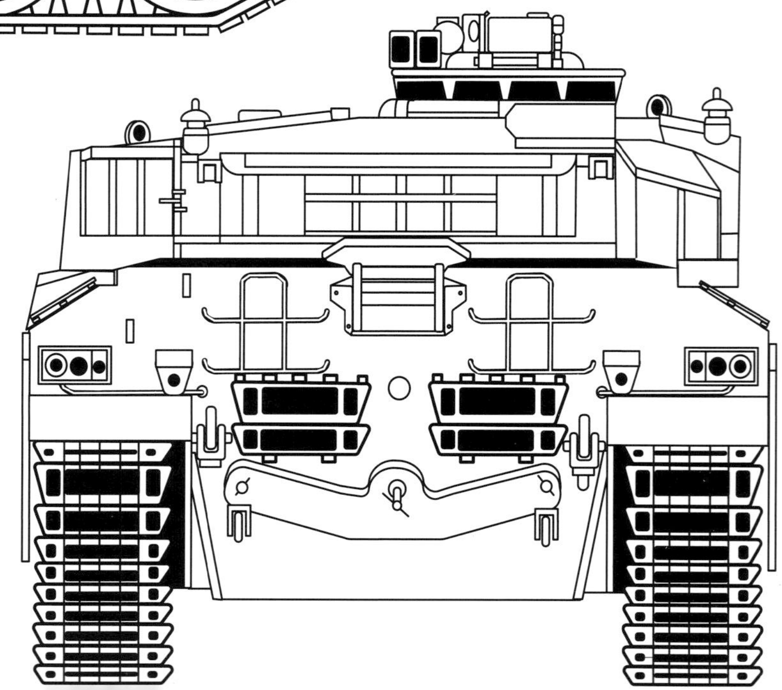

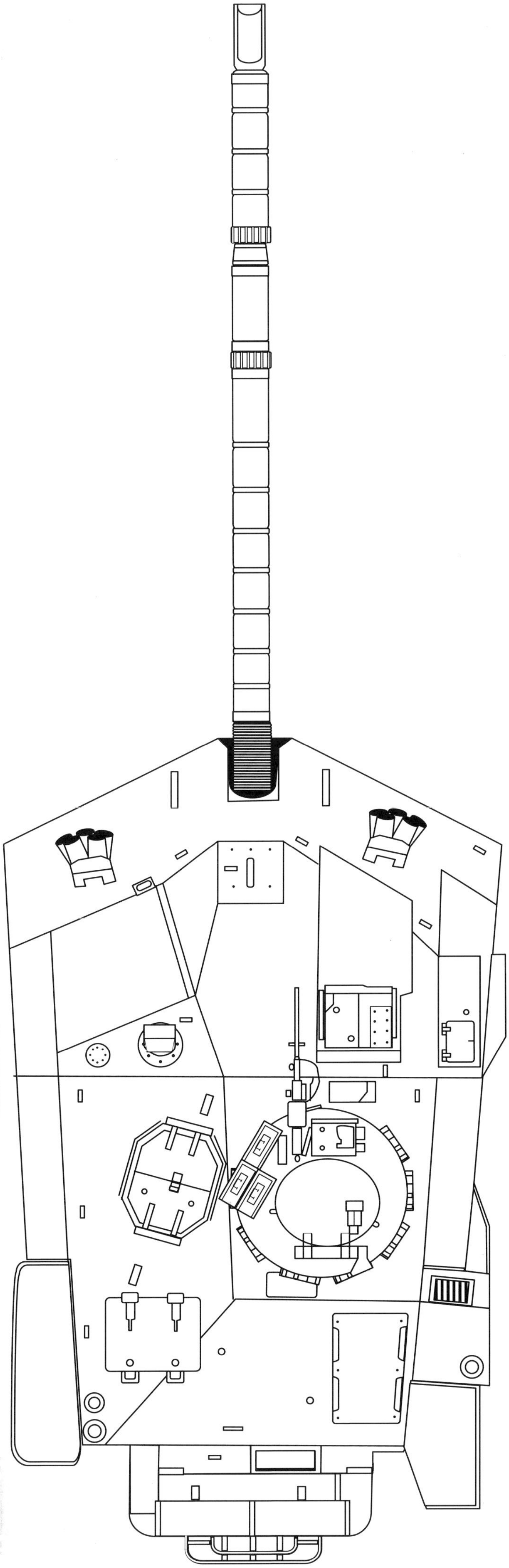

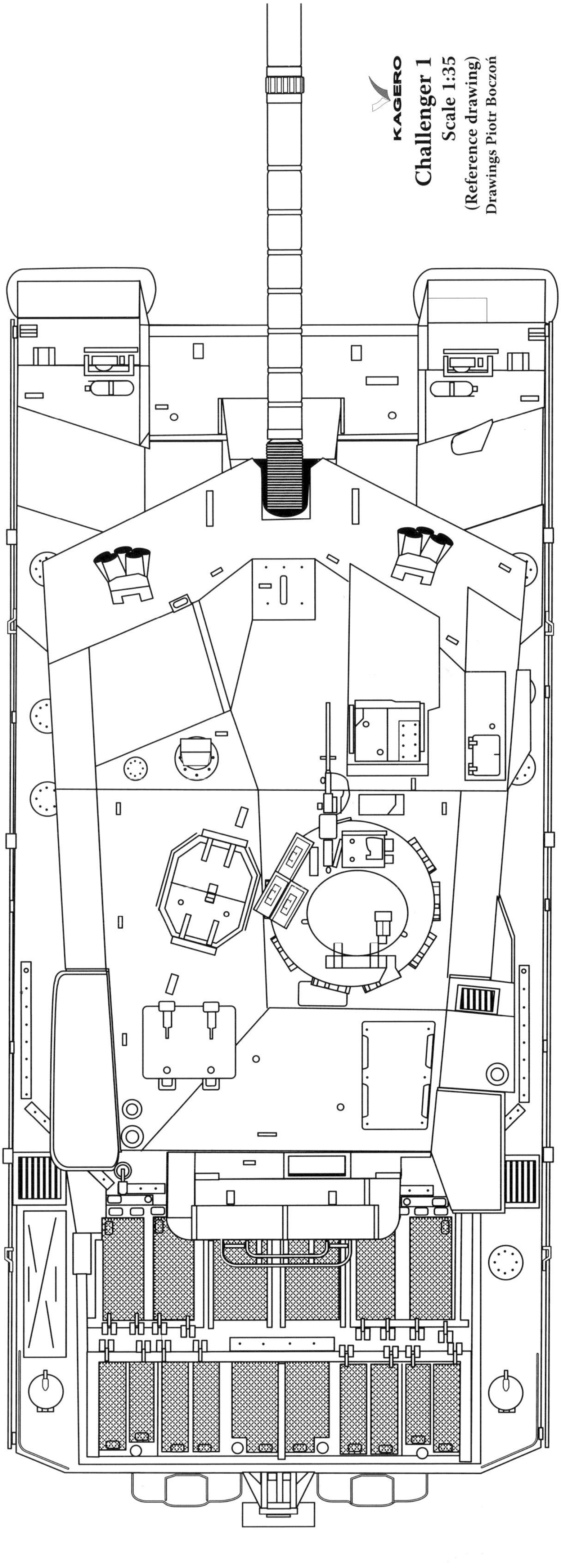

KAGERO
Challenger 1
Scale 1:35
(Reference drawing)
Drawings Piotr Boczoń

In the next volume

Robert Griffin's Challenger 1 Volume 2 follows the Challenger 1's chequered peacetime and ultimately successful combat career. It details events ranging from the disappointment of CAT87 and unsuccessful efforts of Royal Ordnance and Vickers to sell the design overseas, to the type's baptism of fire in Kuwait in 1991. The Challenger's service in BAOR and in NATO's peace keeping missions in the Balkans are also recounted. The story of the CRARRV armoured recovery vehicle and the sale of the gun tanks to Jordan at the turn of the 21st Century round out this well illustrated study. Colour plates covering the Challenger's service in the British and Jordanian armies, a wealth of colour photographs and scale drawings of the vehicle make this an ideal reference.

The Challenger 1 grew out of the Chieftain design through the Iranian Shir program. The Shir 1 design, which had been adopted by the Iranian Army and was on order at the time of the revolution in 1979, became the Jordanian Khalid. It shares a common rear hull design with the Challenger. In these two photos we can see a Khalid being tested with the TOGS system in the 1990s, which in the event was not purchased [D.L.].